ACCESSING THE
INTERNET

A guide for the UK and Ireland

ACCESSING THE
INTERNET
A guide for the UK and Ireland

John Smith
UKOLN, University of Bath

INTERNATIONAL THOMSON PUBLISHING

I(T)P An International Thomson Publishing Company

London • Bonn • Boston • Madrid • Melbourne • Mexico City • New York • Paris • Singapore
Tokyo • Toronto • Albany, NY • Belmont, CA • Cincinnati, OH • Detroit, MI

Accessing the Internet
A guide for the UK and Ireland

International Thomson Publishing
    Commissioning Editor: Samantha Whittaker
    Editorial Assistant: Jonathan Simpson

I(T)P   A division of International Thomson Publishing Inc.
         The ITP logo is a trademark under licence

Made in Logotechnics C.P.C. Ltd., Sheffield
    Project Management: Sandra M. Potestà
    Production: Marco V. Potestà + team
    Artistic Direction: Stefano E. Potestà
    Cover Illustration: William Smith

Printed in the U.K. by Logotechnics C.P.C. Ltd., Sheffield

First printed 1994

International Thomson Publishing
Berkshire House
168–173 High Holborn
London WC1V 7AA

ISBN (ITP UK) 1-850-32147-7

British Library Cataloguing-in-Publication Data
A catalogue record for this book is available from the British Library

Contents

Preface

The purpose of this book is twofold, firstly to give the reader a brief introduction to networking and the Internet, and secondly to provide a detailed list of companies that provide access to the Internet in the United Kingdom and the Republic of Ireland.

Part One starts with an overview of the Internet, defining some of the most important technical terms as it goes. It then describes briefly the basic, common and advanced services or facilities available, with examples and illustrations. It goes on to outline the different ways in which these services are delivered or provided, and the range of connection methods offered.

Part Two lists the companies in the UK and Ireland that provide access to the Internet for organisations and individuals. Full details are given including: name and trading address, range of services offered (including the form of delivery and methods of connection), documentation and support services, the various service packages offered (with current prices), contact information, and some general comments.

Finally, in Part Three, there are some Appendices containing miscellaneous useful information, including lists of the access-providing companies sorted by city where they have local call access, form of service delivery and connection method, plus an extended Glossary explaining in depth many of those abstruse, and sometimes confusing, Internet terms.

The book assumes that readers are familiar, or at least comfortable, with computers, software packages and similar technical ideas. It is a basic guide and not meant to be a textbook or 'how-to' book. There are such books already and a selection is reviewed in Appendix G.

Acknowledgments

I would like to thank the following people for their help in the production of this book:

Chris Brown, Philip Bryant and Lorcan Dempsey (UKOLN, University of Bath), and Eddie Zedlewski (NISS, University of Bath) who read and commented on early versions of the text.

John Stewart (ElectricMail Ltd) who unveiled some of the deeper mysteries of e-mail for me.

Leonard Ainsbury who helped with the initial design of some of the illustrations and especially Figure 3.4.

Chittaranjan Satapathy who sifted carefully through the final draft and made many useful comments on the style and readability of the text for a non-technical reader.

Finally, my special thanks to Samantha Whittaker and her colleagues at International Thomson Publishing who made things happen.

As for the final product – mea culpa.

About the Author

The author is National Project Officer at UKOLN: The Office for Library and Information Networking, based at the University of Bath.

The purpose of UKOLN is to research and promote the use of national and international networks by all forms of libraries (academic, public, industrial, and so on) and information services, and to investigate networks as information access and distribution mechanisms. UKOLN is funded by the British Library Research and Development Department (BLR&DD), and the Joint Information Systems Committee (JISC) of the Higher Education Funding Councils (HEFCs) of the UK.

John Smith's research interests lie mainly in the role of the network as a publishing medium for the academic and research community, especially the possibilities for completely new models of information provision and access. He is also interested in the role of public libraries in a networked information world.

On a wider stage he is concerned with the possible role of the network in the provision of information to the academic and research communities in the developing world.

Part I

The Internet and its Services

Introduction to the Internet

The purpose of this chapter is to provide a very brief overview of networking and the Internet. It is purely a 'scene-setter' for the following chapters.

To begin at the beginning[1] – A potted history of networking

The early wide area computer networks were only available to the military, then the academics and researchers in Universities – notably the ARPANET[2] in the US which began in 1979 and finally closed in 1990, having been replaced by NSFNET. During this period, other US networks came into existence, such as USENET (1979), CSNET and BITNET (both 1981). In the 1970s and early

1980s other countries began to install and use networks, initially for research purposes, for example JANET (Joint Academic Network) in the UK. Initially, many of these networks were based on different underlying techniques and technologies, but there was a steady convergence which finally led to the idea of the Internet (described in the next section).

These early networks were initially meant to provide reliable remote access to scarce resources (as computers were considered before the arrival of the minicomputer and the micro), after which the value of electronic mail (e-mail), file transfer and remote access to distant information services became apparent and the original purpose just one of the supported functions.

In parallel with these developments, the bandwidth (or information carrying capacity) of the underlying technology provided by the national telephone or communications companies was also increasing. It should always be remembered that these networks (in most cases) run over the existing communications infrastructure. There are, after all, no wires (or optical fibres, and so on), in the ground or slung between poles, labelled Internet, JANET, or any other such network.

As the possibilities of e-mail and remote access became more commonly appreciated, commercial organisations wanted to join this new communications medium, but the academic networks, because they were publicly funded, had rules about who could use them and what they could be used for (known as *Acceptable Use Guidelines* or *Acceptable Use Policy*). In response to this demand, the commercial networks came into existence, paralleling the academic networks; using the same protocols, and interworking where appropriate.

The main commercial network operators (for example, Pipex, EUnet, and so on[3]) focus upon selling connections to large organisations who had their own internal networks and so a 'second generation' of network service companies came into existence, buying full connections from the network operators and reselling basic services to smaller companies and the general public.

The majority of companies listed in this Guide are of this 'second generation' type.

What is the Internet?

The Internet is not a network in the simple sense; it is, as its name implies, a collection of networks. What these networks have in common is that they all operate according to certain technical rules or protocols. In the case of the Internet, these protocols are known as the TCP/IP[4] suite. All the networks that form the Internet use TCP/IP, but not all networks that use these protocols are part of the Internet, as some are private and un-connected to the Internet. The fact that all the member networks use the same underlying protocols enables the other defining factor of the Internet – its parts or members have agreed to interwork.

Currently there are approximately ten thousand networks forming the Internet. This consists of over one million computers, each having unique address or Internet number, with over 20 million users. The reasons for the vagueness in these numbers are twofold.

1. Because the Internet is a network of networks and these top-level networks may themselves be composed of yet smaller, separately governed, networks, it is impossible to know the exact numbers.

2. The growth rate, estimated at over 10% per month, means that any sensible estimate is out-of-date by the time it is printed![5]

Why connect?

You may think this is an impressive piece of technology – but what can you do with it? There is, in fact, a range of services provided by the Internet and we will look at the operation of these in greater detail in Chapter 2. Here we will quickly review some uses.

 I cannot emphasise too strongly that the following paragraphs represent only a tiny fraction of the possibilities of the Internet. A list of further reading is available in Appendix G.

Business use

E-mail is becoming a recognised form of communication for a wide range of users. BBC Radio 4's Today programme now gives out an e-mail address with its paper mail address and fax number when asking for its listeners' comments. As an example, when I recently organised a four day conference on networking, I sent and/or received over one thousand e-mail notes, sent and/or received nearly two hundred faxes, but did not send or receive a single letter!

You can use the Internet for international, as well as national e-mail, because using Internet e-mail costs no more to the other side of the world than to the next town. The expression 'global economy' is being used increasingly to indicate that no aspect of business or commerce operates in isolation and that developments in distant parts of the world can have an effect locally. The growth in the use of e-mail and the Internet will serve to accelerate these changes. Already systems engineers in Ireland maintain computer systems in the US by accessing them remotely across the Internet. India is becoming a major player in the world's software industry, stimulated, at least in part, by the presence of the Internet. As the technological prophet Arthur C. Clarke has demonstrated by conducting his business worldwide from his home in Sri Lanka, where you live need not necessarily affect what you do for a living.

E-mail is also ideal for collaboration. The speed of a telephone call combined with the ability to store the ongoing 'conversation' means that e-mail is better than letter, fax or telephone in the support of collaborative working. We are all aware that the financial world never sleeps, with one stock market opening as another closes, a pattern which is repeating itself in other areas of work. In another example from the area of computing, software is being written on a continuous 24 hour basis with programmers in the UK passing on their day's work to programmers in the US, who in turn pass on their work to others in the Far East. The updated work is passed back to the UK the following day. Each person involved works a normal day – no night shift. There must be other areas where such concentrated activity could give a company or organisation a competitive edge.

Using the basic medium of e-mail (or a facility known as USENET News) you can participate in hundreds or thousands of ongoing electronic discussions. This could be for business, recreational or educational purposes. In the business area you can access the latest share quotes, company telephone numbers, discussions of up-and-coming technologies, and so on. At present most of these discussion groups are free to access. As the Internet becomes more widely used, more and more commercial 'value-added' subscription services will be made available, especially in the areas of technology support, information location and finance.

Educational use

In schools, e-mail is being used to enable children to communicate with others around the world. This not only improves their knowledge of other cultures but also their ability to use the written word.

Using more sophisticated tools, children are reaching around the world and picking up satellite weather maps, working with hundreds or thousands of others on environmental projects, or even improving their foreign language skills.

The same is available to adult learners with access to hardware, costing less than a thousand pounds. Experimenting with a recent software promotion distributed free in a computer magazine, and with access to a free trial dial-up service, I was able to reach sources in America, Russia, and China from rural Wiltshire.

On a more prosaic note, public libraries in the UK are beginning to explore the possibilities of using Internet to make local information available on the networks, something their counterparts in the US have been doing for several years. Academic libraries in the developed world have had computer based catalogues (also known as OPACs – On-line Public Access Catalogues) for many years; thousands of these are now connected to the Internet, making them available for direct searching from all around the world.

Organisations like the Open University have been experimenting with networking for the support of distance learning for some time, using the academic networks. The steady growth in the availability of the Internet to non-academic users will only serve to accelerate this process.

Recreational Use

The world wide e-mail lists (or e-mail conferences) and USENET News carries thousands of discussion groups concerned with every recreation known to man. For example there are discussions on golf, tropical fish, where to get the cheapest air tickets to Tahiti, computing problems, evaluation of software packages, philosophy, music (from heavy metal to Indian classical), and so on. At the moment there are tens of thousands of discussions going on, with topics as diverse as AIDS and Zoroastrianism. You can either find a recipe for tomorrow's lunch or ask a world expert a question on artificial intelligence with equal ease.

Obviously education and recreation are closely intertwined. If you learn something new about your hobby or sport, it not only improves the enjoyment of your recreation, but also increases your knowledge.

Packet Switching and Client/Server – Two technical concepts explained

Although you do not need to understand the technology of the Internet to use it, two technical concepts are so important that it is worth giving some time to explain them. These two ideas are *packet switching* and the *client/server* model. The first underpins the operation of the network and the second underpins the design of the interactive services offered on the network.

Packet switching

The Internet is what is known as a packet switched network[6]. This means that data is broken up into smaller chunks or packets to which address information is added. These packets are then sent across the network until they reach their target where the address information is removed and the original message or file is reconstructed. The usual analogy used to explain this concept is to think of a letter or packet which consists of, an envelope or wrapping which bears the address, and the contents (data). These packets are passed between the computers (or nodes) that form the infrastructure of the network until they reach their destination. Each node reads the address on the 'envelope' and routes it on towards its destination according to the state of the network at that time (hence these nodes are called *routers*). However, the state of the network is constantly changing (machines may be busy, lines congested, and so on) and this means that the packets that form a particular message or file will not necessarily follow the same path across the network. For this reason the network is often depicted as a cloud with machines connected to it, but with no clear connection between individual machines. In Figure 1.1, the PC at bottom left is communicating with the large microcomputer at top right and the Mac is communicating with the mini at top left. The packets from each machine enter the 'cloud' and follow the route that is both available and appropriate at that time. This is a very simplified diagram as there is usually other hardware between the user machines and the network, sometimes a LAN and at the very least a router. With a basic access service (from work or home) there is likely to be at least two modems and a dial-up telephone line between the end-user and the access provider's machine. It is then likely that between the access providers machine and the Internet, there will be a router and possibly a security machine. The configuration at the far end could be equally as complex, though if it is a service, the modems and dial-up line will be replaced by a leased line, router and LAN. It should also be emphasised that the diagram assumes an IP connection[7] to each computer shown. Other forms of end user access are described in Chapter 3.

The details of the operation of packet switching are not important, but the implications are that any such network is inherently robust, flexible and efficient. It is robust and flexible because if any part of the network fails or is congested the computers at the nodes (the *routers*) simply find another path around the blockage and the network continues to operate, though its speed of response may change. It is efficient, because high-speed lines are shared by many users simultaneously, rather than being dedicated to one user at a time as, for example, telephone lines. In fact the Internet could not exist in any other form - it is the advantages of packet switching that make it possible.

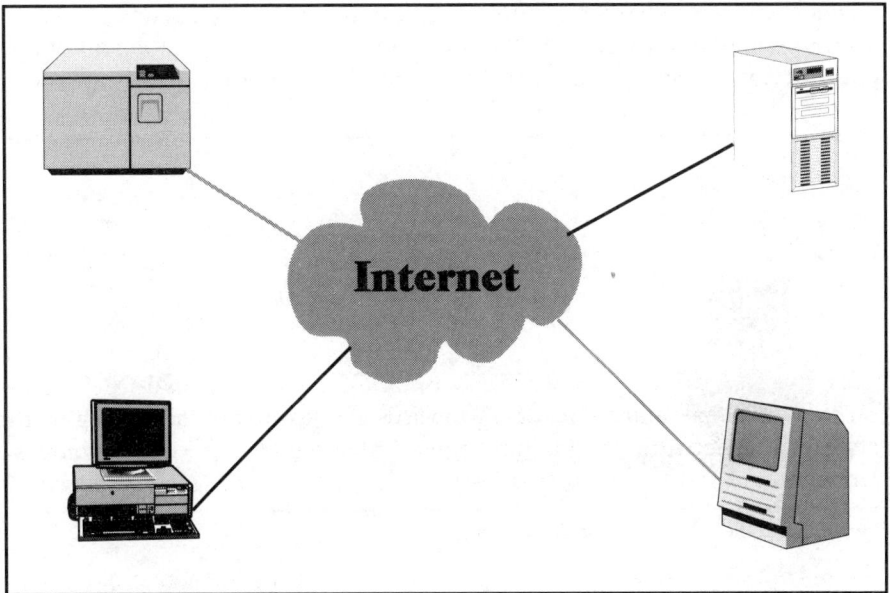

Figure 1.1 Network 'cloud'

The client/server model

You will come across this phrase so often when reading anything about Internet interactive network services, that it needs to be explained early on.

In order to complement the flexibility offered by packet-switching, and support the open interworking of machines from different manufacturers (which is a fundamental part of the philosophy behind TCP/IP), the interactive services offered over the Internet needed to have an equally flexible design. The approach adopted is the *client/server* model. This means that instead of having one program on one machine working to provide a service like *remote access* (*telnet*) which requires an interaction between two machines, there are two complementary programs (one on each machine) that cooperate to achieve the desired result. The usual rule is that the program started by the user on the local machine which requests a service from a distant machine is called the *client*, and the program that provides the service is called the *server*. Again, because the system is designed to accommodate any machine, the two programs need to communicate in a standard form known as a *protocol*[8]. The functionality of the two cooperating programs and the protocol is exactly defined in the related standards.

This is illustrated in Figure 1.2. The zig-zag line represents any communications carrier medium. It could be a simple serial cable between two machines in the same room, a LAN, an international network like the Internet or a combination of these and other elements. The major requirement of the link is that it is able to support the required standard protocol.

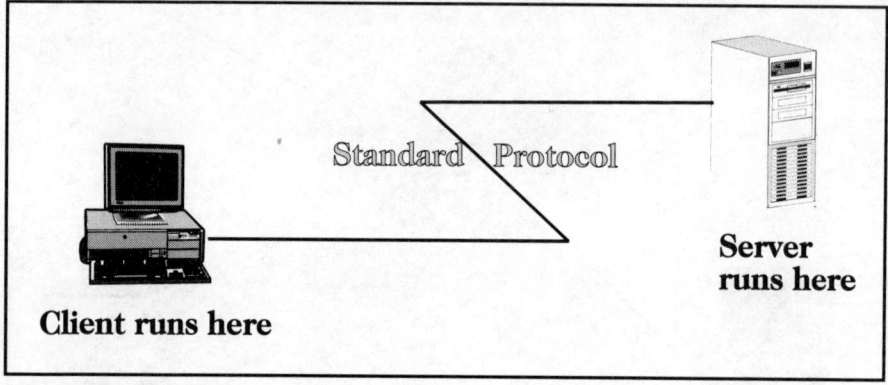

Figure 1.2   Client/server – basic model

In summary – what this means for a user is that when you start a service, for example, *telnet* (which enables remote access to distant machines), what you are doing is starting a program on your local machine which interacts with a complementary program on the distant machine in order to achieve the desired service (remote access).

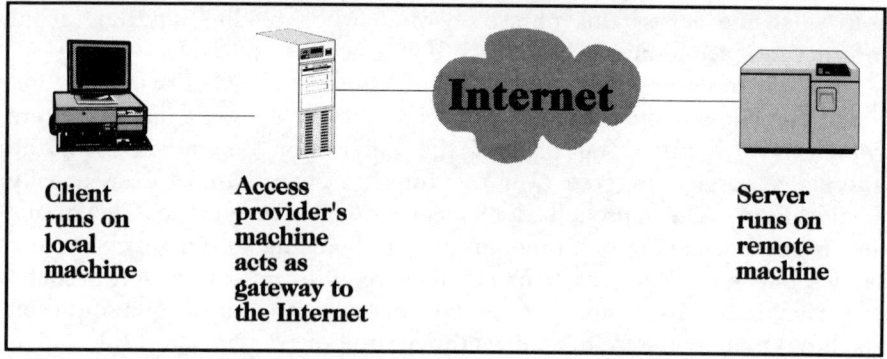

Figure 1.3   Client software on local machine

If you are connecting direct to the network from your own machine (with all the operational programs running on your machine), this clean separation of the client and the server programs and the standard protocol that enables them to work together also means that you can buy whichever client package you prefer, or suits your environment. The make and model of the server package is irrelevant.

If you opt for terminal access, there is a slight complication as the client programs now run on your access provider's machine, not your local one. You have a connection to these programs via the terminal emulator program running on your local machine. The difference is clearly shown in the illustrations (Figures 1.3 and 1.4). There are other options for accessing the network – where the programs run, and what this implies, is discussed in detail in Chapter 3.

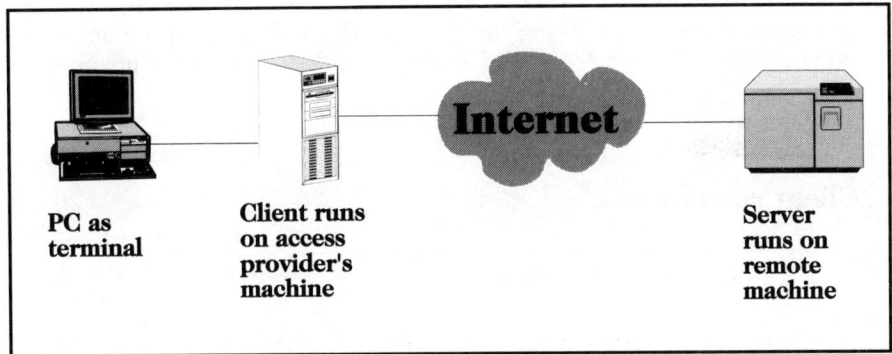

PC as terminal   **Client runs on access provider's machine**   **Server runs on remote machine**

Figure 1.4   Client on access provider's machine

Finally – and most importantly – the way a client/server based service looks and behaves depends entirely on the client package you are using: the server has no control over this. Therefore the same information from the same server may look very different for users with different clients. This will be illustrated in Chapter 2 when we look at some of the advanced services based on this model.

Notes

1   With apologies to Dylan Thomas!

2   An expansion of this acronym and the other network names on this page and elsewhere can be found in the expanded glossary.

3   Pipex and EUnet are the main public access IP network providers in the UK at the time of writing but the very large telecommunications companies are waking up to the growth of the Internet. British Telecom and possibly others (for example, Sprint International, Mercury Communications, and so on) may be providing IP services by the end of 1994. *See* Appendix B for further information.

4   Transmission Control Protocol/Internet Protocol.

5 Another statistic that brings home the growth rate of the Internet is: In 1984 there were 1 000 machines connected, by 1987, 10 000, by 1989, 100 000, by late 1992, 1 million.

6 The Internet is not the only packet switched network and TCP/IP is not the only packet switching protocol. X.25 networks are also packet switched networks.

7 This means the data packets originate on, and are delivered to, the machines involved.

8 The protocols required by the basic services, e-mail, telnet, and so on, form part of the TCP/IP suite of protocols, the protocols that enable the operation of the newer services (Gopher, WAIS, and so on) assume the underlying presence of these basic protocols).

2

A review of Internet services

This chapter is concerned with the services supported by the Internet.

The exact package of services provided by an access provider will vary. All access providers offer e-mail and the majority offer the basic services described in the following section. A growing number also offer a selection of the other services described later in this chapter.

The services are divided into *basic*, *common* and *advanced*. The logic behind this division is more historical than real – the basic services came first and are defined as part of the Internet protocols, the common services are just that, common but not part of the core Internet services (and in some cases do not depend on the presence of the Internet), and the advanced services are the latest developments. NB: advanced does not mean difficult to use or only for advanced users – the opposite is often the case, with the latest services often being the easiest to use.

For each service covered, there is a description of what it is used for and how it works.

Basic Internet services

Although there are many new tools and services becoming available for exploring and using the Internet, at heart it provides three basic services: Electronic mail (e-mail), File Transfer, and Remote Access.

Electronic mail or e-mail

E-mail is a one-to-one or one-to-many medium. It can be considered as an electronic analogy of paper mail (often called 'snail-mail' by networkers to indicate its comparative slowness!). However, in addition to speed of delivery, e-mail has other advantages over paper mail. It is very easy to send copies to multiple recipients. This may be done by including all their addresses in the note passed to the mailer program or by keeping groups of recipients' addresses in what is known as an 'e-mail address list' (or e-mail list). When you need to send a message to a pre-defined group you mail it to the list rather than to each recipient individually.

The apparently simple idea of having a separate list of e-mail addresses has become very important outside of personal e-mail. By moving such lists into a public area and making it possible for recipients to join or leave the list at their request, the idea of the e-mail conference or discussion list was born. Using this idea, millions of people are able to join in electronic discussions on thousands of topics. Another common use of discussion lists is for one user to ask for specialist advice concerning the topic of the list, thus tapping into the knowledge of tens, hundreds or even thousands of fellow readers. For some of the lists, especially those concerned with the maintenance of computer hardware or software, this 'advice seeking' may be the most common use.

Another advantage of private e-mail is the ease with which you can add comments to an incoming note and then forward it back to the sender, or on to someone else for further comments or to be actioned. This can be useful for a developing discussion between two or more users, as the history of the discussion can be encapsulated in the note[1].

Because e-mail is so pervasive the following description is more detailed than is the case for the other services.

The most basic form of e-mail consists of a file of text in a standard format. The details of how the text (assuming it is text) is produced and how it is viewed depend on the local mailer software[2]. The format of any e-mail file consists of a *header* and a *body*. The header includes (but is not limited to) information about the sender, the recipient, and the subject of the e-mail. The body contains the actual message

Before a note can be read, it has to be sent. The note is usually 'composed' using an editor, which may be either part of the mailer package or an external editor. The great majority of mailer packages prompt the user for basic information (recipient name/address and subject), although only the recipient name/address is absolutely essential. You are then presented with an empty 'body' in which to compose your text (or copy ready-prepared text into it). The mailer package then puts the completed note in a queue to be transferred to the recipient's mailbox. In its simplest form, e-mail may simply involve users on the same machine – in which case the network is not needed. Assuming it is addressed to a user on a distant machine, it will need to be passed from the sender's machine to the mail service on the distant machine for onward transfer to the recipient. If the distant machine is busy it may require several tries before the mail is finally delivered. On the receiving machine it is stored in and initially read from a common area. It may be moved to the reader's disk-space if he or she decides to keep a copy for later use. Although e-mail can often appear to be instant from sender to receiver, the design of the system allows for situations where this is not possible, for example when either of the machines is busy, the network is congested, the distant machine is 'down' for routine service or repair, and so on. In any of these cases the sending machine will leave the note in the queue and try again later. It will continue to do this for hours (or in some systems, days). If it proves impossible to deliver the note within the preset period, it will be returned to the sender with a message saying that it could not be delivered and possibly including a suggestion as to what the problem may be.

Once a note has been delivered to the recipients' machine, they need to read it. To do this, a local mailer package is started and used to read the incoming mail. Because the form of e-mail is standardised, it can be composed using one package on one machine and read by another package on a completely different machine.

Although the way e-mail looks depends on the program used to read it, the following sequence of illustrations[3] (Figures 2.1a, b and c) show what it is like to start a typical mail reader (in this case, one called MSG on a UNIX machine) and read a note.

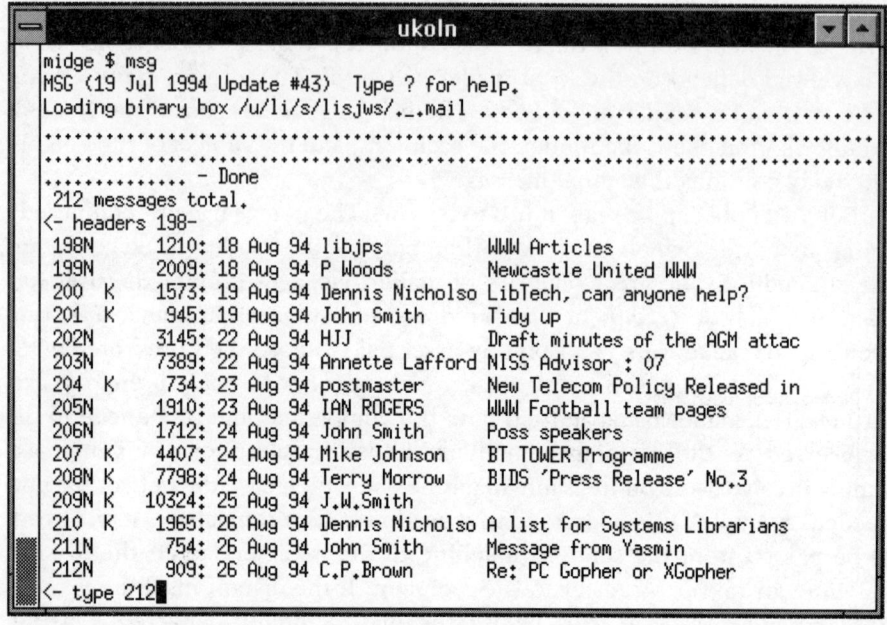

Figure 2.1a   Initial list of most recent notes

The information in Figure 2.1a is taken from the headers of the incoming notes and presented as a summary of the contents of the user's e-mail in-tray. Having selected a note to be read (or in this case 'typed'), we see Figure 2.1b.

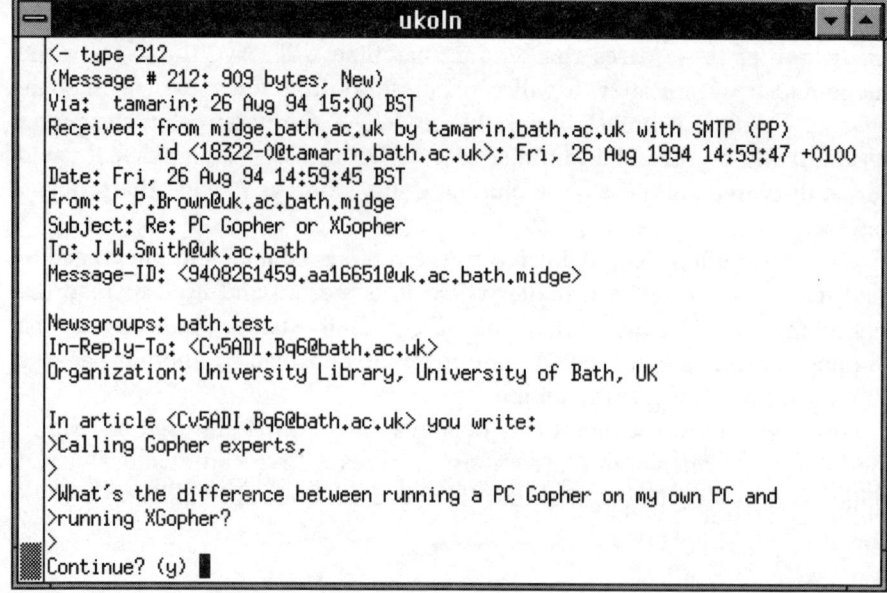

Figure 2.1b   First half of chosen note

In Figure 2.1b the portion of the note from 'via...' to the first blank line is the *header*. Just like the postcode on paper mail, or the related dots added to envelopes by the Post Office for their automatic sorting machines, you don't need to understand much of this to use e-mail. The basic details – From, To, Date, Subject – are obvious. The From and Date fields are filled in by your mail package when you send a note. In this case the note is longer than the screen so we need to type 'y' to read the next screenful.

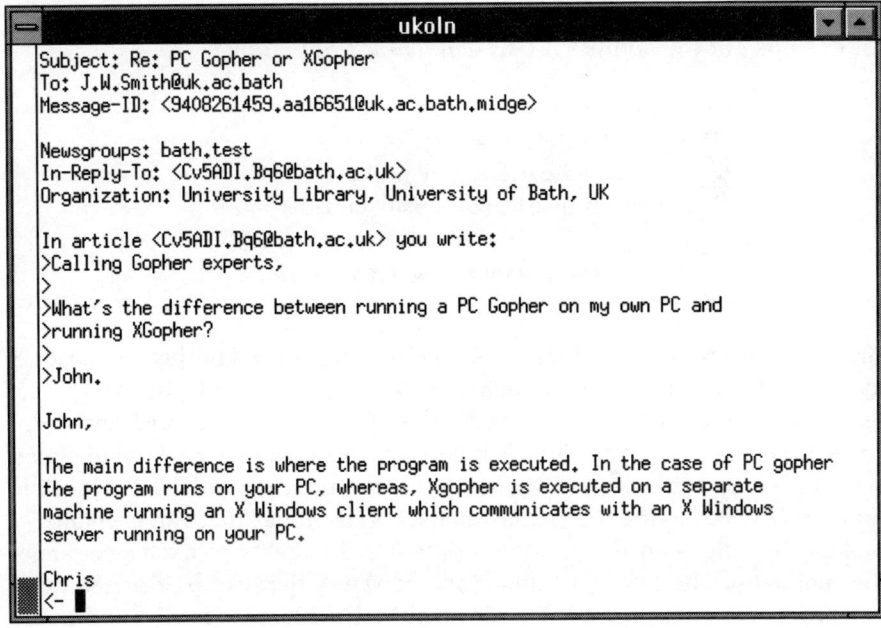

Figure 2.1c   The rest of the note

For many years this 'text-only' form of e-mail was the only form. Now there are newer forms which allow the inclusion of non-text information (for example, digitally encoded graphics, sound, computer programs in binary form) in a mail message. The standard that allows this for Internet e-mail is known as MIME (Multipurpose Internet Mail Extensions). Although mailer packages that support this standard are becoming more common, you cannot assume that every recipient will have access to this, so it is worth checking before trying to send someone e-mail that includes binary information using this standard.

Because e-mail files transferred between machines are stored in a common area, the sender only needs to know the recipient's e-mail 'address', which does not include a password. An e-mail address consists of the recipient's *userid* (or a standardised form of their real name) and their machine's network address. The usual format for Internet e-mail addresses is:

---

username@site.network.countrycode

---

or, for sites with more than one machine:

---

username@machine.site.network.countrycode

For example, in the address *A.User@machine1.bath.ac.uk:*

A.User	is the user's name
machine1	is the user's mail machine
bath	is the site (University of Bath)
ac	is the UK academic network (JANET)
uk	is the countrycode for the UK[4].

There are variations. For example, as the Internet started in the US, most US addresses do not include their countrycode but instead end with the type of network the addressee is on: 'edu' for educational, 'com' for commercial, and so on.

Because text-only e-mail is so simple, it is usually quite easy to send e-mail to recipients on networks that are not part of the Internet. Indeed it is possible to find references to the Internet as that set of machines that can exchange e-mail easily. This is an incorrect description – the ability to exchange e-mail does not define the Internet. Among the networks that can be reached by e-mail from the Internet are: BITNET, EARN, UUCPnet, FidoNET and many X.400[5] commercial e-mail services. The ease with which you can reach users on these networks may depend on how the service you subscribe to handles such transfers. In many cases it will be easy and completely transparent, but with others you may need to be quite knowledgeable about e-mail addressing systems. If you are looking for an access provider, and mail exchange with non-Internet sites is expected to be part of normal use, it would be sensible to check how easy this is before subscribing.

File Transfer

This allows for the transfer of any computer file between two machines. It is usually used to transfer larger text files (many e-mail systems have limits on the size of a note) or for non-text files. Such files may contain programs in binary form, or be the data files of common applications like word-processor or spreadsheet packages. It is often forgotten that such files contain layout and processing information in addition to the textual or numeric information, and if any attempt is made to transfer them as text files (in their original form) they will be irreparably damaged and the delivered files rendered unusable. Although the contents of the file to be transferred may be in any format and may be text or binary, whether it is ASCII or binary does need to be specified to the file transfer software before the transfer takes place.

The protocol that defines how this is done is called, the File Transfer Protocol (FTP) and the program that does the transfer the File Transfer Program (again FTP). In the case of File Transfer (unlike e-mail) it is possible to instigate the transfer from either the sender's or the recipient's end. Also, when sending, the file is delivered to the recipient's private disk-space, and when retrieving, it is read from the distant user's disk-space. For these reasons, basic FTP transfers require the instigator of the transfer to know the userid, machine address and password of the recipient if sending, or the userid, machine address and password of the provider if getting a file from elsewhere[6].

However, there is an easy to use form of FTP, known as *anonymous FTP,* where userids and passwords are not required. In this case, files are made available in public areas of some networked machines' disk-space. It is then possible, by linking to such a machine using the FTP program with the public userid *anonymous,* to explore the public directories and retrieve files. When using the *anonymous* userid you are still prompted to enter a password, and the convention is that you enter your e-mail address as the password[7]. Although *anonymous FTP* is a simple idea it has enabled the provision of gigabytes (possibly terabytes[8]) of textual and binary information, spread over 1000 – 2000 machines on the Internet. A machine that allows *anonymous FTP* access is said to have (or be) an *FTP archive.* Some machines are dedicated to the provision of such services.

Remote Access (telnet)

This refers to the ability to connect to a distant machine across the network. You may be simply logging in to your own machine from a remote site, or accessing a remote service, for example a database host service.

The protocol and program that enables this is called *telnet.* As with FTP, the word *telnet* may be used both as a noun and a verb, that is, when using the telnet facility you are *telneting.* In most cases you will need a logon account for the remote machine, unless it allows public access, usually by using some form of 'guest' account.

Common services

Bulletin Boards (BBs)

The concept of the Bulletin Board (BB) is similar to the idea of a public Notice Board in the paper world, where messages may be 'pinned up' for others to read. As with the public Notice Board, it is expected that the messages are relevant to those reading the board regularly. The use of BBs is very similar to the use of discussion lists described above under 'E-mail', and the term 'bulletin board' is sometimes used to refer to discussion lists.

Many BBs are offered on small PCs attached to a dial-up modem and have nothing to do with the Internet. Commercial Internet access providers obviously run larger machines and when they offer a BB service it usually covers tens or hundreds of topics. In some cases the information on a 'board' may have some commercial value, so you need to subscribe specifically to that service and pay an extra fee for accessing it. In CompuServe, access to the local BBs (known as Forums) is one of the major reasons (along with the access to commercial and non-commercial databases) to join.

USENET News

USENET News is a form of distributed Bulletin Board (BB). It pre-dates the public availability of the Internet (and many other data networks) and was designed to operate efficiently in a world where machines called each other over ordinary dial-up telephone services and did so at relatively infrequent intervals, for example, once a day.

In use, News[9] looks very similar to a local BB service and the basic software can also be used to provide such a service. Each individual BB is known as a 'Newsgroup' and you subscribe to the Newsgroups that cover the subjects you are interested in. However, where the local service provides access to the national or global News services, things change. Whereas a local BB service may have tens or hundreds of BBs, News consists of approximately 5000 Newsgroups (though few sites provide access to all of them). Local users read and write (in News terms, 'post') to local and global Newsgroups in the same way. USENET is based on a hierarchical structure where large machines provide a *Newsfeed* to smaller sites (which in turn may provide a feed to smaller sites, and so on). Periodically the local system will wake up, collect together all the new local postings to national or global Newsgroups, and connect to the machine above in the hierarchy, passing up the new local material and collecting any new material from outside. This new material is then added to the local pool of News information. The local system may also do the same for machines it feeds, or wait until they contact it. The machine 'above' will pass this new material (combined with material from other 'lower' machines) up to the machines above it, and down to other machines or across to peer machines. Each 'posting' has a unique identifier indicating where and when it originated. It is worth remembering that, because each item posted propagates across the network a little like a wave spreading out from a pebble dropped in a pond, users at different parts of the network can look at their local version of a Newsgroup at the same time and see slightly different contents. Older material will be common to all the sites but newer material will appear first at the sites nearest to the site where it was posted.

As indicated above, very few ordinary sites take all the Newsgroups, yet the *average* multi-user News site receives over 20 Mbytes a day (a full feed is over 50 Mbytes/day and rising). To put this in perspective – 20 Mbytes is approximately equal to 2.5 million words! Obviously only the very largest sites would have space for more than a few weeks' worth of such traffic, so News incorporates a rolling deletion facility. Depending on the perceived importance of a particular Newsgroup, the items in it may be kept for a few days or for a month or so – this is defined by the site for each Newsgroup taken. Knowing that this has to be the case, to keep the thing within manageable proportions, some items are updated and re-posted (and therefore re-broadcast) regularly so they never disappear from any local service. Usually these items are concerned with the operation of USENET News itself or are introductory files of information for new users (like those found in the Newsgroup 'news.announce.newusers', for example).

If some of the numbers above seem a little daunting, remember that if you join a USENET service you only see those Newsgroups you choose to see. This is one of the advantages of both BBs and News over e-mail discussion lists. If you subscribe to a list, a copy of all the mail sent to that list will be sent to your mailbox (something to remember if you go away for a long holiday!), but with News only one copy of each message is kept on each machine that takes a particular Newsgroup (no matter how many users of that machine subscribe to that Newsgroup), and that copy is not on your disk-space. Further, older items will be periodically thrown away to make space for newer items, even if no one reads them.

In order to read News you need a *news reader* program. In most cases this runs on the access provider's machine. The most common readers in the academic world are *rn (read news)* or its grown-up brother *trn (threaded read news)*. The preferred option is *trn* because it sorts incoming articles within Newsgroups into related groups or threads. A typical example of *trn* use is given in Figures 2.2a, b, c and d[10]. Another option is to have a News reader program on your local machine. You can either read the News interactively off the access provider's machine in real time (that is, you will need to be logged in while you are reading), or the latest updates to the Newsgroups you have previously selected can be downloaded in a batch when you log in, or at predefined intervals, and you read them off-line later.

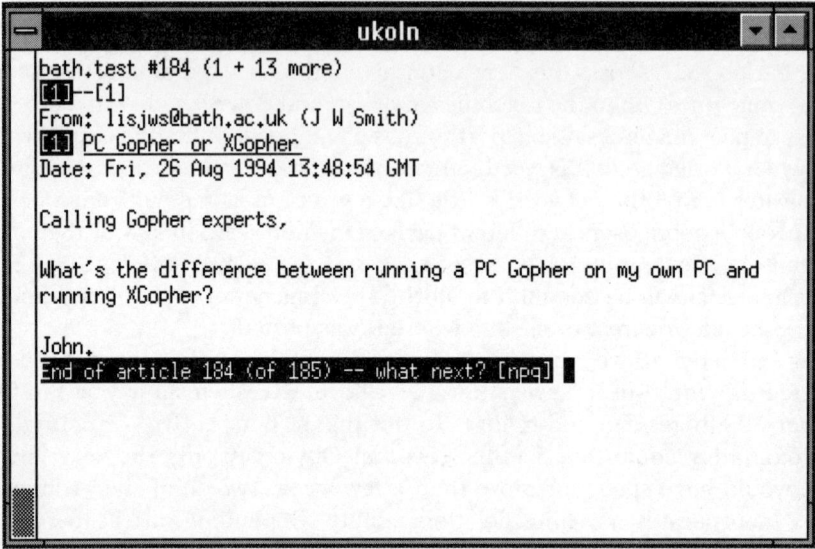

Figure 2.2a A sample posting

In order to set up this sequence of screens and demonstrate *trn,* the question shown in Figure 2.2a was posted in a local test Newsgroup. This was considered safer[11] than using one of the real Newsgroups concerned with Gopher or X-Window development, as they have worldwide distribution.

```
midge $
midge $ trn
Unread news in bath.general                         68 articles
Unread news in bit.listserv.lis-l                    4 articles
Unread news in news.announce.conferences          173 articles
Unread news in news.announce.newusers              41 articles
Unread news in news.announce.newgroups             67 articles
etc.

======  68 unread articles in bath.general -- read now? [+ynq]
======   4 unread articles in bit.listserv.lis-l -- read now? [+ynq]
====== 173 unread articles in news.announce.conferences -- read now? [+ynq]

======  41 unread articles in news.announce.newusers -- read now? [+ynq]
======  67 unread articles in news.announce.newgroups -- read now? [+ynq]
====== 604 unread articles in news.newusers.questions -- read now? [+ynq]
======  65 unread articles in uk.announce -- read now? [+ynq]
======  59 unread articles in uk.events -- read now? [+ynq]
====== 163 unread articles in bath.changes -- read now? [+ynq]
======  25 unread articles in bath.test -- read now? [+ynq]
```

Figure 2.2b Starting *trn*

As the local version of *trn* starts it indicates the status of the top five of the user's chosen Newsgroups. At this point, if there are any new Newsgroups their names are listed and the user is asked if he or she wants to subscribe; in this case there are none. Then, one line at a time, *trn* displays the status of the Newsgroups previously chosen and in the preferred order. This is known as the Newsgroup selection level of *trn*. At position 10 the local Newsgroup 'bath.test' is displayed and selected.

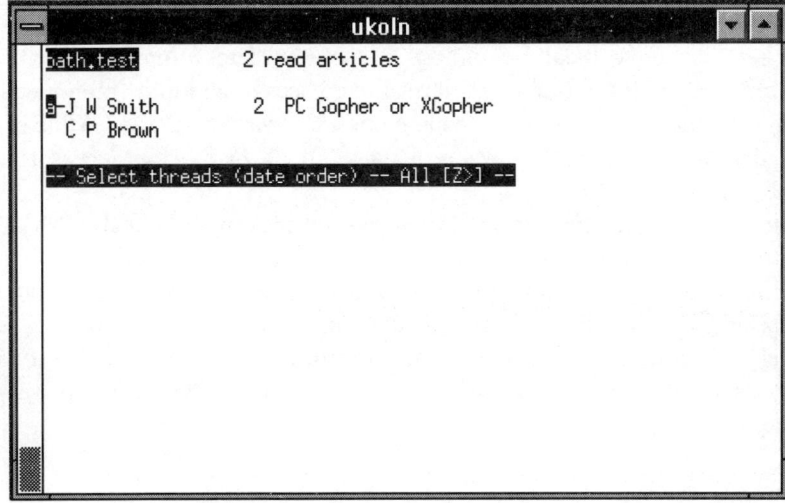

Figure 2.2c   The thread level

In Figure 2.2c the thread consisiting of the original query and a reply is shown. There were other threads in the Newsgroup bu these have been hidden to simplify the display.

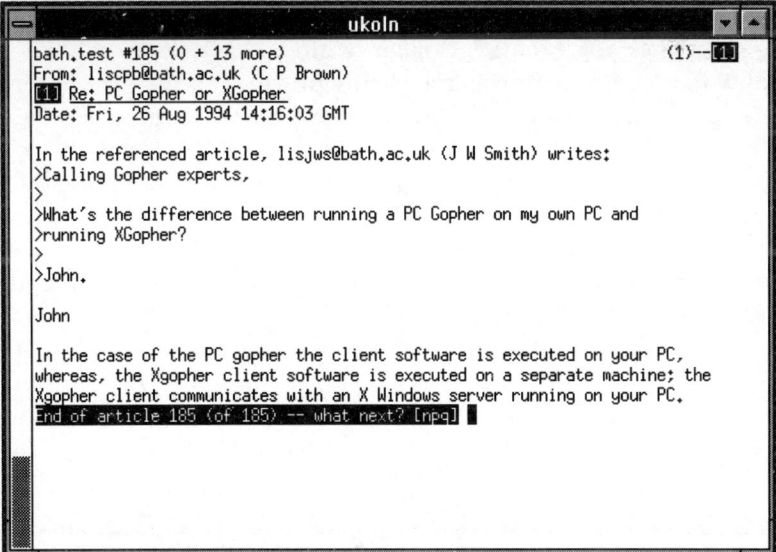

Figure 2.2d   The article level

Figure 2.2d shows the reply. Here the person replying has used one of the features of *trn* (and most other News readers) to embed the original article in the reply. This is useful for others who may be following this discussion or who may have a similar query.

Databases

Some services have locally mounted databases. These may be databases containing local information, commercial databases of all kinds (bibliographic, holidays, financial, and so on), or collections of software packages. While many of the databases are free, there is usually an extra fee for access to the commercial ones.

The Poptel service (and the related Host services in Manchester, Kirklees, and other places) also offer access to remote commercial databases via a local interface. This local interface interacts with the user, helps formulate the query, logs on to the distant service, does the search, and then logs off, thus keeping the additional access costs to a minimum. A neat idea if you can't justify a subscription or you don't want to learn a range of different search techniques.

Advanced services

All these services are concerned with finding and retrieving information scattered around the Internet and for this reason they are often called *Networked Information Discovery and Retrieval (NIDR)* tools.

The main ones are: Archie[12], Gopher, WAIS and WWW (World Wide Web or Web). Where a service provider explicitly provides access to such services they are listed in the *Additional services* section of the provider's page in Part Two.

The following short sections outline the main features of some of the more advanced Internet information discovery and retrieval tools. These are brief descriptions, not detailed 'how to use' instructions.

The purpose of these sections is to give a 'feel' for what tools are available and what you can do with them.

Archie

Archie is an access tool for locating files in archives accessible using anonymous FTP as described in the section on File Transfer above.

The major problem with anonymous FTP is that there is too much out there. There are currently over 1500 anonymous FTP sites and before Archie became available you either had to link to each one and search its directories for the desired file, consult a networked or printed subject directory or ask a friend. Now, Archie short-circuits all this.

One of the commands available in anonymous FTP allows you to ask the distant server to return a listing of the contents of all of its directories. You can do this manually, but visiting each server in turn, entering the command and then scanning the output would be impossible in any realistic amount of time. The Archie program automates this procedure. It contacts all the anonymous FTP sites registered with the Archie service and collects a listing of the titles (filenames) of all the public files stored. It then merges all this information as a textual database. Every unique filename is listed, together with all sites that have a copy. In addition, the database notes the size of the file and the date of the last update. There are approximately 25 Archie servers around the world and they all contain the whole database. The purpose behind this replication is to allow users to consult the one closest to them in order to minimize network load. Also, rather than all 25 servers regularly searching the Internet (which itself would impose a large load on the network, the archive hosts and the servers), the database maintainers cooperate and each Archie server collects from a designated area and then shares the data found with the others. It must be emphasised that what is stored is the *name,* not the *contents,* of the file (even if the contents are text, which in the majority of cases they are not). This means it is only useful if the managers of the FTP archive sites choose to give the stored files meaningful names. Some FTP archives are designed to be used with their own indexes, so their files have names like `technical_report_243` or, even worse, `2d_27.Z`. All these will be indexed by Archie, but unless the user has some insight into the meaning of these filenames, they are useless as search keys.

Because of the variability of filenames and the possibility that they do not reflect, in any meaningful way, what they contain, a secondary database is also available. This is the 'description database'. When files are added to an FTP server, the providers may also supply Archie with some descriptive text which is stored in a separate database. The command to use this is 'whatis *search-string*'. The search-string is compared with the text record in this separate database and Archie gives you a list of files, each with a short descriptive string. If one or more of the files look relevant you need to go back and perform a standard Archie search, looking for the filenames found by 'whatis'.

There are three ways of using the Archie service:

1. Run an Archie client on a local machine. On a UNIX machine this is just like using FTP or telnet. All the commands and the text-string to search for are entered on the command line as arguments (or parameters). A list of filenames and sites matching the search-string is returned. Although this is the most efficient way to use Archie in

terms of network and server use, and your time, it lacks some of the options available when using e-mail or direct access (as described below) – for example, the 'what is' command is not available.

2. By e-mail. In this case the commands and search-string are sent in the body of an e-mail note to one of the Archie servers. In the same way, the result of the search is returned to the original sender's address (or to another e-mail address) as an e-mail note.

3. By direct telnet connection to an Archie server. In this case you log in as 'archie' and interact with the Archie program at the distant site. This last option is not recommended unless you have no local client and e-mail from your site is very, very, slow. It ties up network and server resources, and the Archie server user interface is as intuitively obvious as a Chinese proverb written in Greek. If you must use it, the commands are similar to those used in the e-mail version.[13]

The Archie server for the UK is at Imperial College (archie.doc.ic.ac.uk). For start-up information send a note with the word 'help' in the text body to 'archie@archie.doc.ic.ac.uk'. If you have a signature appended to the bottom of all your e-mail, put the word 'quit' on the next line after the word 'help', or Archie will try to read your signature, looking for commands – in which case the results may be unpredictable![14]

Internet Gopher

Gopher is a menu-based package that provides a simple hierarchic front-end (or user interface) to a range of services on the Internet. It is a client/server system, that is, you interact with a Gopher client program running on your PC or Mac (if you have TCP/IP direct to the desktop) or on the host service you log into to access the network, and it interacts with a distant[15] Gopher server which contains information or pointers to information or other services. There are currently over 1400 Gopher servers. Gopher can also start up other Internet services (like telnet) on your behalf, which makes it a very flexible and powerful tool. Gopher servers are used to provide local information services like Bulletin Boards, Campus Wide Information Services (CWISs – pronounced Kwiss or Seawiss) and access to distant information services. Many governmental and inter-governmental organisations are putting up Gopher servers with information about themselves and their operations. For example, the World Bank has a Gopher server (gopher.worldbank.org), and so does the World Health Organisation (gopher.who.ch).[16]

As a Gopher client starts up, its first move is to contact its designated default server. The address of a particular server is initially built into the client software, but it can be changed once the program is running or overridden at start-up by giving the address of a server as an argument (or parameter) to the start-up command.

Without the connection to a server the client would be like a person without a personality – form without content. Though, as is usual in computing, there is an exception even to this rule – see the discussion of the 'hotlist' below.

Once started and connected to the server, the client displays a menu of options. A range of possible options is described in the list below. The text following the hyphen refers to the layout of a VT100 based Gopher interface (like that shown in Figure 2.3). The screens displayed by PC and Mac based clients use more easily understood graphic icons to indicate the type of the item. The two illustrations below the list show just how different a menu from the same server can look when accessed using two different clients.

Basic options in a Gopher menu:

- A text file – indicated by **.** at the end of the line.

- Another menu – indicated by **/**.

- An interactive service (usually a keyword search engine – indicated by **<?>**). If you follow this you will be prompted for the text-string to be searched for.

- A graphics file – indicated by `<picture>`.

- A binary file that can be downloaded – indicated by `<bin>`.

- A telnet connection to another service – indicated by `<telnet>`.

- A sound file – indicated by `<)`.

There are others, but like the graphics and sound options above, you may only be able to view or hear them if you have the appropriate local hardware and software.

One of the options within all Gopher clients is to make a note of sources you have found useful so you can return to them at will. This list is known as your 'hotlist'. In this case you do not need to rely on a connection to a remote server before you can connect to your favourite sources – you can select them from what is, in effect, your own private Gopher server (though it contains only pointers, not information).

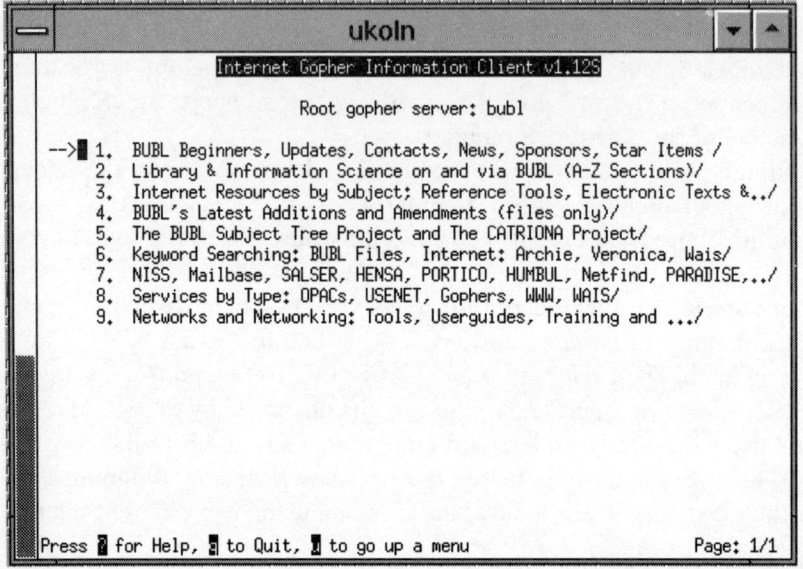

Figure 2.3 View of BUBL[17] Gopher server main menu using a VT100 client

Because of the power of the Gopher model and the way in which much of the operation is hidden from the user, some common misconceptions have arisen about how it operates and these can cause confusion when there are problems. The most common misconception is that you 'link-through' from one server to another. It is not unusual to hear a new (or experienced) Gopher user say something like "I linked to the Birmingham Gopher which connected me to the Texas Gopher which linked me to the local state library catalogue". What actually happened is that the user's local Gopher client first connected to the Birmingham Gopher server. In the item the user selected from the menu offered by Birmingham was the address of the Texas Gopher; the local client then used that address to connect to the Texas Gopher and displayed its menu. The user then selected from that menu an item that contained the telnet address of the state library's on-line catalogue. Again, the local client took that information and made a telnet call to that address.

The reason it appears that you are being 'linked-through' is because the local client 'remembers' where it was last and when the current operation is closed (that is, you choose to move back to the previous menu) it remakes the link to that service. This is like reading a book with lots of cross–references and keeping a finger in one page while going to another page to read a note or comment and then coming back to the original page – except that what is really happening is that you are making a note of the number of the page and not really keeping a connection to it. The purpose of this is to make the network appear to be a seamless navigable space (known as Gopherspace). The implications of the real, rather than the apparent, underlying model of the Gopher system are:

- You only communicate with one server at a time and that connection only exists for as long as it takes to send a request for a menu, file, or whatever, and receive the result.

- It is possible that when you attempt to move back to a previous menu item, you cannot re-establish the connection to the appropriate server – at this point the seamless model of Gopherspace is shown to be apparent rather than real.

- The services that you can reach may be limited by the connectivity of the machine on which the client is running; for example, if the client is running on a machine that has been deliberately limited to be able to link only to local machines (for reasons of security, for example) then despite the fact that a local Gopher server can point you to distant services you cannot connect to them.

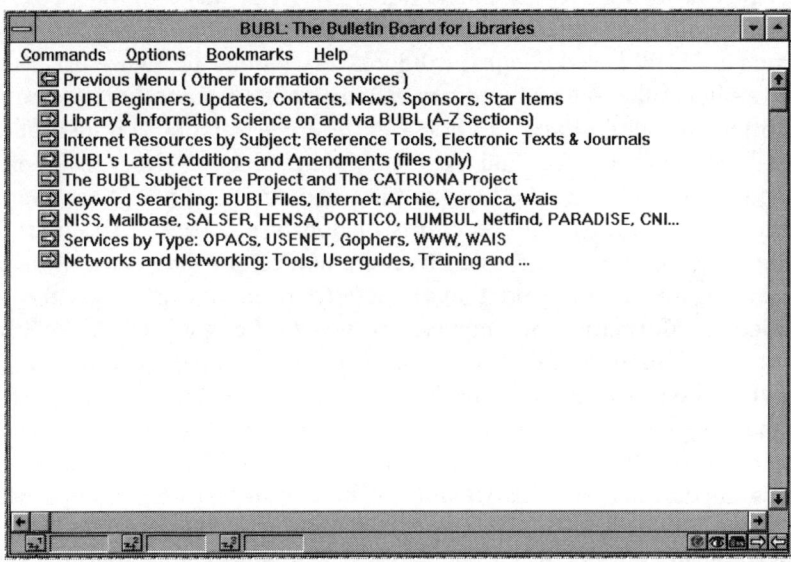

Figure 2.4   View of BUBL Gopher server main menu using a Windows based Gopher client

Any one of these pieces of information or services may be on the local machine or on a remote machine. The beauty of Gopher is that the user does not need to know where it is or how to get to it.

Exactly how you interact with a Gopher client will depend on the one you are using. With the simple VT100 client interface shown above, you use the cursor control or arrow-keys to move the 'arrow', [Enter] or the [→] key to select an item, [Spacebar] to move to the next screen of a multi-screen menu, and so on. With a PC Windows or Mac based client you would do all this and more with the mouse.

The simplicity and basic user-friendliness of the Gopher model is also one of its limitations. When you first link to a Gopher server you can see only its top-level menu – you cannot see what else it has without exploring each of its menu items in turn, which may lead to yet more levels of menu. This was acceptable when Gophers were used mainly for local CWISs or similar services, but it does not scale up to a worldwide service where a menu item may point to a menu on another server, each item of which is a pointer to yet another menu elsewhere, and so on. The solution was to have a service a little like Archie (*see* above) which collects together the text from all the known Gopher menus and makes them available in a keyword-searchable form. This facility is called *veronica*[18]. Not all Gopher servers have a veronica service but most have a pointer to one. In many cases where the server itself does not have this service it may offer a choice of other servers that do. All other things being equal, you should use the one geographically closest to you in order to minimise net traffic. The way a veronica service looks when you connect to it will depend on the client you are using. If you are using a VT100 based client it will look very similar to Figure 2.7, as this is just Gopher collecting an input text string on behalf of another application. In addition to being offered a choice of veronica servers you also have a choice of what is searched. You may search a database of the titles of either all menu items including the titles of files, and so on, or just those that are the titles of directories. The second is obviously a subset of the first. The advantages of searching only the directory names are that you will usually get a much smaller list and that, hopefully, each one is a pointer to a collection of information on your chosen topic. The result of any search is just another Gopher menu, only in this case it is one made just for you. You may follow the links and come back to it, just like any other menu.

Some Gopher servers also offer a local keyword search service (for example, BUBL offers such a service) which gives access to the full-text of the files on that server. This should not be confused with veronica which gives access to directory and/or file names only. With one of these local full-text search services you are usually accessing a WAIS service (*see* below) in the background, but the result of the search is presented as a Gopher menu. Figure 2.7 illustrates the input stage of using one of these services.

WAIS – Wide Area Information Server

Like the other major information access tools, WAIS is based on the client/server model. Basically a WAIS server provides a database of fully indexed full-text documents. It may also contain files of non-text material but these will need some text for the system to search on.

WAIS client software is available for most popular personal computers and workstations and many of these packages are available on the network via anonymous FTP. Most service providers that provide TCP/IP to the desktop will also have copies of shareware and freeware WAIS client packages. Many services that provide interactive access from the host computer will have a VT100 based WAIS client available. The following illustrations (Figures 2.5 and 2.6) show examples of WAIS clients. The first is an example of an X-Windows WAIS client (running on a UNIX system but viewed using X-Windows from a PC) using the NISS[19] WAIS server, and the second is a VT100 based client. The second is also an example of a client made available remotely to enable users without their own local client to access a WAIS server. In this case it is the WAIS client made available by NISS as part of the NISS Gateway service.

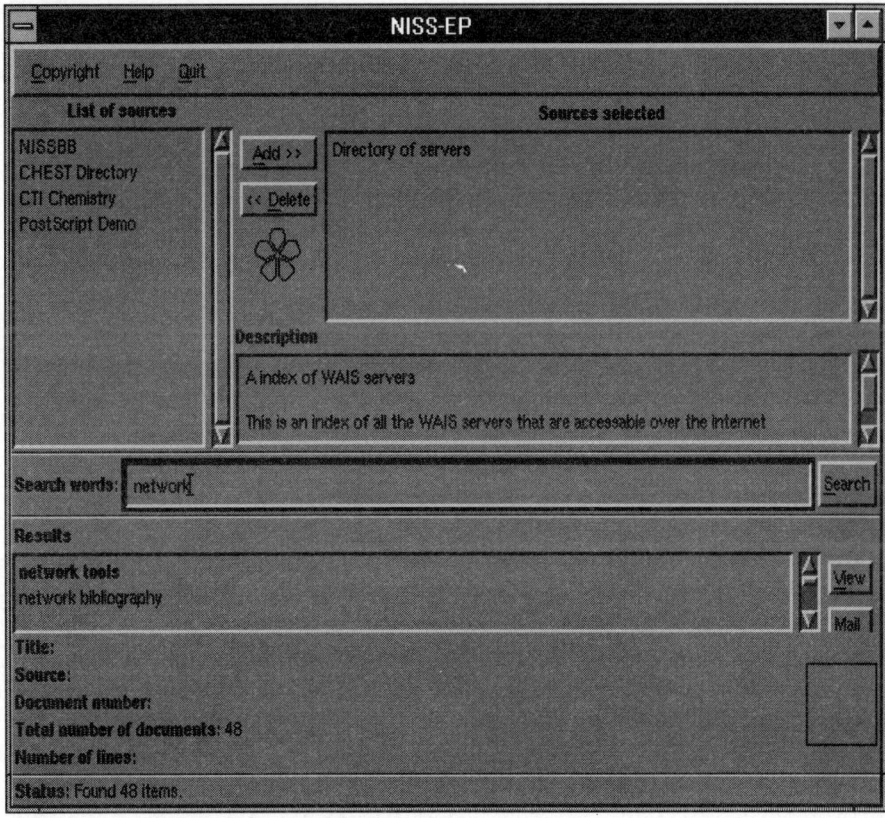

Figure 2.5 X-Windows based WAIS client

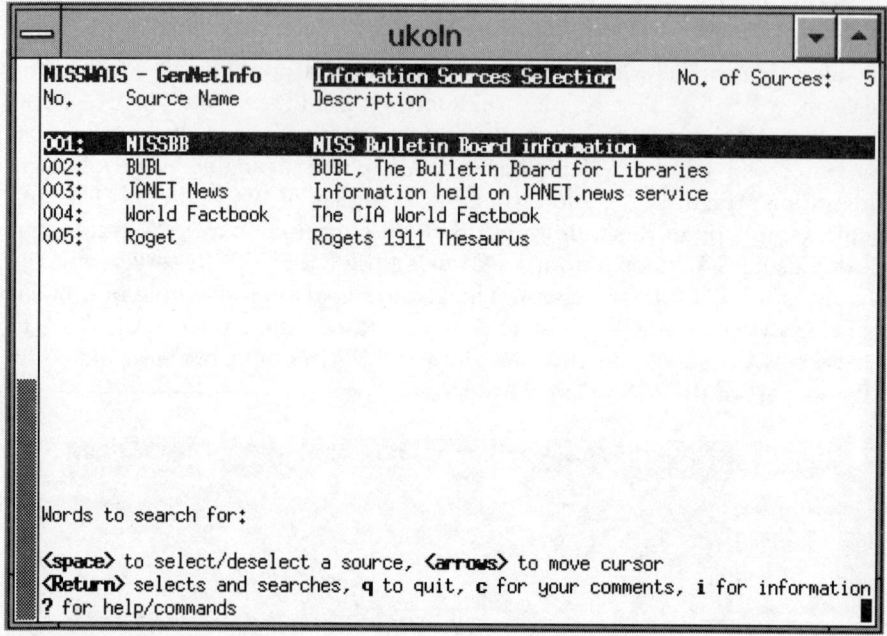

Figure 2.6   VT100 based WAIS client

It is quite possible to be linked to a WAIS server via a Gopher client. In this case some of the limitations indicated below will apply, but you will not be warned you are about to do a WAIS based search and the results may look odd unless you are aware of this. Also, the Gopher interface to external services like WAIS is so limited that you cannot use most of the good features of WAIS. An example of what you would see using a VT100 based Gopher client to link to a WAIS system is shown in Figure 2.7. In this case the Gopher client is simply collecting the search-string on behalf of the WAIS system and then displaying the result in Gopher form. Although this hides most of the complexity of a true WAIS client from the end user, it also hides most of its useful features.

Each WAIS server site contains a collection of indexed documents. These may be simple text or other forms of information (for example, graphics images) but with some indexed text to enable their location and identification.

For those familiar with 'conventional' free-text search database systems, WAIS seems a little odd. The basic idea of WAIS is that the indexing software on the server indexes all the words in the documents, including those which would normally be considered 'stop-words', that is, words with no retrieval value like 'and' or 'the'. Secondly, it responds with a list of documents in ranked order, the most relevant (according to the WAIS algorithm) first. Thirdly, it incorporates 'relevance feedback', which enables a user having

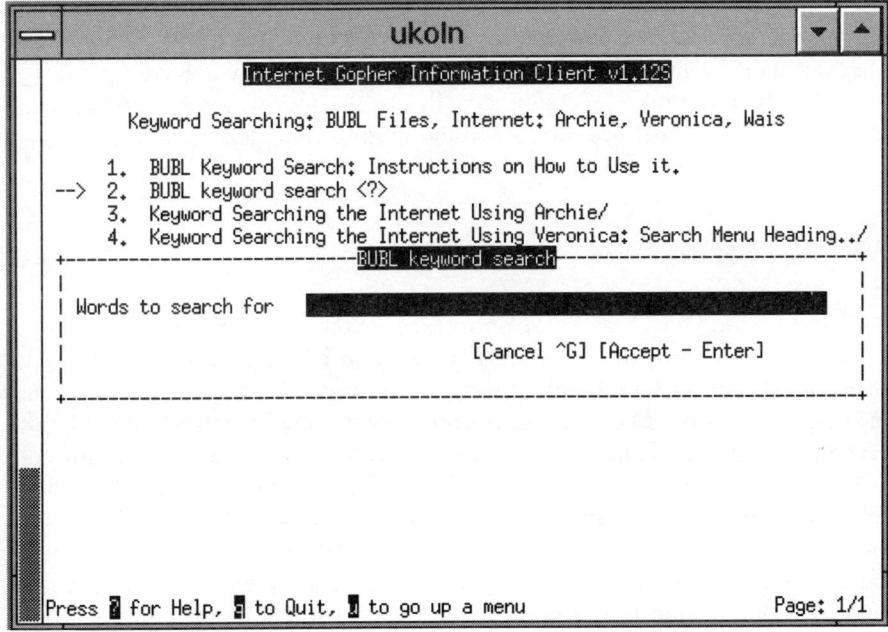

Figure 2.7 VT100 Gopher with 'input box' open

found a useful document, to request the system to 'find more like this one'. Although the user's view of WAIS will depend on the WAIS client being used, the basic functionality and operation are the same. Assuming you are using WAIS for the first time, the client will start off only knowing about the servers whose addresses are 'hardwired' into it, one of which will be the 'directory of servers'.

Assuming you do not know the best source (server) for the type of information you want, you need to search the 'directory of servers' first. To do this select the 'directory...' as the source to be searched and enter appropriate keywords for the search-string (remember that at this time you are searching short text documents describing source sites, so use general terms). The result will be a list of sources. Most of these sources are free but a few will be subscription only, so you can see that they may be relevant but you cannot access them[20]. You then choose one or more that look useful and enter the same, or more specific, keywords. The result this time is a list of documents. You can then choose to look at these documents. You can 'flag' those which seem most relevant and then feed these back when asking for 'more like these'. The exact way in which you carry out these activities will obviously depend on the client used. WAIS is much, much easier to use with a real graphics interface complete with windows and a mouse, though it can be used successfully via a full-screen terminal and cursor control keys.

If you are connected to a WAIS server from a Gopher client (as discussed above), the sources will already have been selected (by the Gopher server designer) as relevant to the topic being searched for, and you have no control over this. In addition, the 'relevance feedback' option for locating further documents based on those already found is not available.

WWW, World Wide Web or Web

The World Wide Web (Web is the short form of the name, WWW takes longer to say than the words it stands for!) is based on hypertext. In its most basic form this is text with embedded pointers to other text that expands on the current text, so instead of a linear document with one beginning and end you have many documents intertwined. One example (which requires no hi-tech to use) is dictionaries that use a different font to indicate words used in definitions that are themselves explained elsewhere in the dictionary. Another example often seen in modern magazine articles and text-books is the use of a separate 'box' or 'sidebar' where topics are explained or covered in greater detail than in the main text. Many CD-ROM based encyclopedias use hypertext links from text to text, or text to pictures, and so on.

The Web uses the same idea but now the links may point to other parts of the same document or to another document, or part of a document, on a local or distant server. At the core of the Web is a standard form of 'mark-up' language (HTML – Hyper Text Markup Language) which is used in the creation of the source documents.

Figure 2.8 shows a PC Mosaic Web client looking at the home page of the University of Bath's Web server. Where text is underlined, it indicates a hypertext link to another document. The fact that the underlines are dotted means that the link has been followed before by this client.

It is possible to have line-based Web clients, but they are very painful to use and lose so much of the conceptual richness of the hypertext idea. The very least you should use is a full screen character based client like Lynx. This is a VT100 based client that looks a little like a Gopher client, but supports the hypertext links found within Web documents. As you can see in the following illustration, where graphics are displayed by Mosaic or Cello or other graphics-capable clients (*see* Figure 2.8) when they are available from a Web server, Lynx can only indicate where a graphic would be if you could display it.

Like Gopher and WAIS, all Web clients need to have a pre-selected server from which to start; this is known as the 'home page'. Again, like Gopher and WAIS, this starting point can be changed. Once started, it is difficult to describe what happens next, since if you are browsing[21] a hypertext document you can branch to another part of that document or to another document by selecting one of the highlighted words.

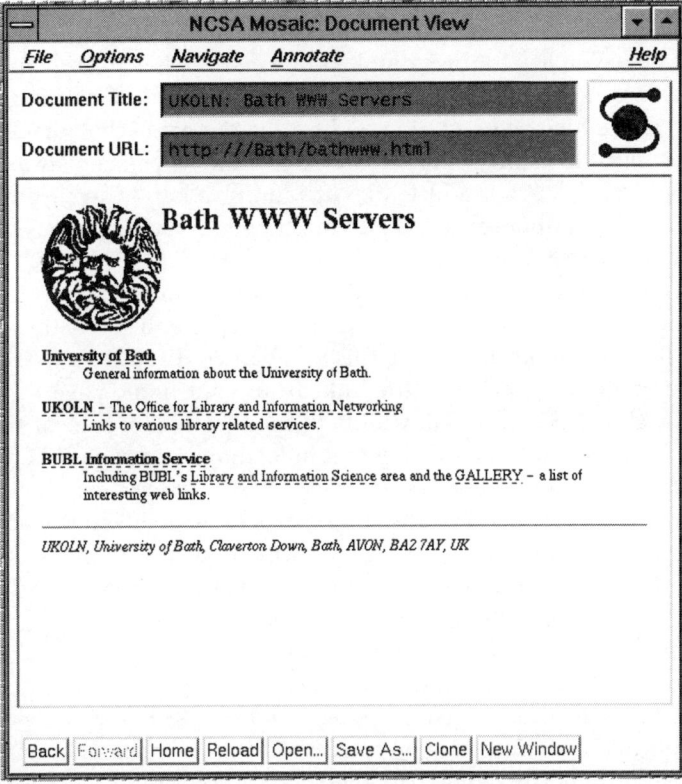

Figure 2.8 PC Mosaic Web client

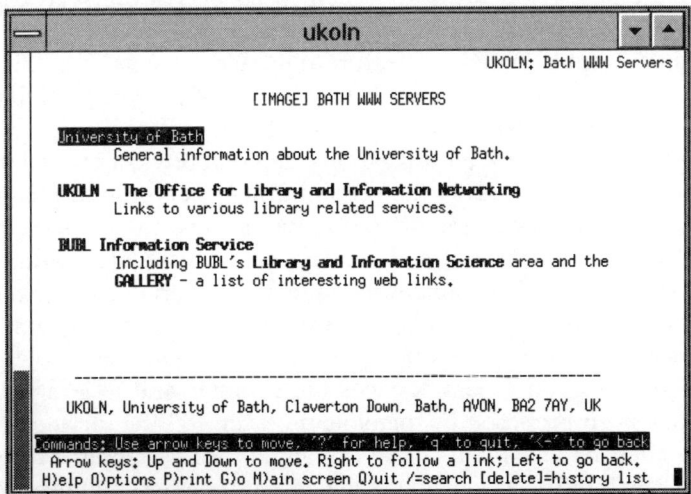

Figure 2.9 Lynx VT100 Client looking at the Bath home page

There is no need to return to a menu or list of possibilities. Indeed, one of the major problems in using the Web is keeping a clear picture in your own mind of where you are in this complex information-space. To help with this, all Web clients keep track of where you are in terms of the page you are accessing and a history of where you have been during the current session. All clients offer the facility to move back up a link, so you can always re-trace your steps. This feature is a little like unwinding a ball of string as you enter a maze, so that, no matter how lost you may get, you can always find your way back to the beginning.

Like Gopher clients, Web clients also have a 'hotlist' facility where you can store the addresses (or URLs, *see* next paragraph for details) of pages that you have found useful. Using this you can go directly to those pages in the future without following all the links from your home page.

Because of the power and flexibility of the Web idea it is clear that, in the same way that Gopher can overlay and hide simpler services like telnet and FTP, a Web client can access Gopher and FTP servers as a less flexible subset of itself. For this reason a Web client needs to know in advance just what sort of server you are asking it to link to, so it will know how to make the link. To enable this there is an expanded form of network address known as a URL (Uniform Resource Locator), which includes this detail. For example, an FTP address in URL form will start `ftp://` and a Gopher address `gopher://`. By default, the Web client assumes that any server you point it at is a Web server. Figure 2.10 shows what a Gopher server looks like when viewed via a PC Mosaic Web client. As you can see, the Gopher menu items are considered as hypertext links (indicated by underlining in this version of Mosaic). This is the same Gopher server top-level menu as displayed in Figures 2.3 and 2.4, yet again the client used makes the server look very different.

The more sophisticated or 'feature-rich' Web clients (like Mosaic and Cello) that run on workstations or high-powered personal computers can handle text, graphics, sound or even moving images[22].

The only problem with all this power and flexibility is that you must read true hypertext documents on the screen, otherwise you lose much of their impact and/or information content. You cannot print out a sound (though you may be able to print out a picture if you have the appropriate support software and printer). You can download text, for example, but if you print it out you may find it has references to information (via hypertext links) which is not there in the paper version.

Because the Web can access services like Gopher and FTP as a subset of itself, it has been proposed by many as the over-arching service into which the others will be absorbed. There are already some access providers (for example, Pavilion Internet) who assume their customers will be using a Web client on their home machine. These companies (and others) also provide space on their Web servers for customer use.

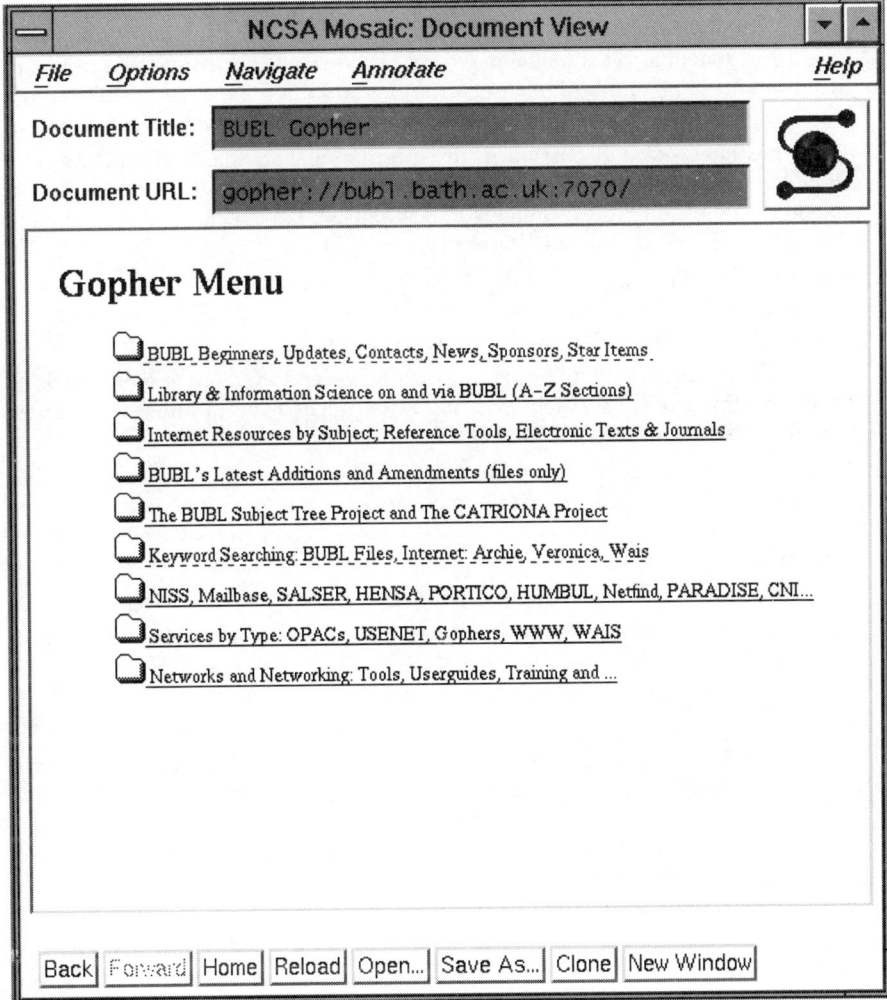

Figure 2.10 Mosaic Web client view of a Gopher server

Notes

1 Although this feature is very useful it should not be misused – otherwise there may be multi-page notes flying around the network with only one line of new information in them. This is waste of bandwidth and bad *netiquette*. For an explanation of *netiquette* see the expanded Glossary.

2 Also referred to as the Mail User Agent.

3   These three screen shots are of an *X-Windows* session running under Windows on a PC, communicating over an Ethernet LAN with a mailer package running on a local UNIX machine. This model is typical of the way users gain access to e-mail in universities and similar large organisations (except for the use of X-Windows which is less common). It is a form of terminal access, as described in Chapter 3.

4   There has been some discussion in the international standards groups that this should be 'gb' (as is used in X.400 e-mail addresses). However, the code 'uk' is now so commonly used that it is unlikely to be changed, but 'gb' may, at some time in the future, become an official alternative.

5   X.400 is the OSI e-mail standard. Most large international commercial e-mail providers offer X.400 services rather than the Internet form. A list of main X.400 service providers in the UK can be found in Appendix D.

6   A rarely used option even allows the instigator to send from one remote machine to another by proxy, in which case he or she might need to know the userids, machine addresses and passwords of both of the other parties.

7   Most systems don't mind if you don't, or if you enter some nonsense string, but others get quite annoyed and threaten not to let you in next time you access the service if you don't give your real e-mail address. Computers have no sense of humour!

8   Gigabyte=10^9 bytes (1000 Megabytes), Terabyte=10^{12} bytes (1000 Gigabytes).

9   Although for the rest of this section I will use the short form 'News' instead of 'USENET News', it should not be confused with the other News service, found on many multi-user machines, which is used to keep users up to date with local system or service developments.

10 As with the e-mail examples, these screen shots are of an X-Windows session onto a local UNIX machine. *trn* would look similar when used via any VT100 (or better) terminal emulator. NB: As with most of these packages, there are many options set in the background that can affect the detailed way *trn* looks and acts. Therefore this sequence can only be seen as typical of *trn,* not definitive – it could look different in detail on your access provider's machine.

11 Strictly speaking, it is possible to post an article to a worldwide Newsgroup and limit its distribution to local machines. However, one slip of the finger and you can send an unintended message to thousands of readers! Sending test articles out onto the whole net is considered bad *netiquette.* Also, using a local test Newsgroup gives more control over any other contents.

12 It must be confessed that Archie is not as popular, or as common, as the other three (Gopher, WAIS and WWW) – nonetheless it is a very useful tool and therefore worth describing in some detail.

13 It is probably truer to say that the e-mail commands resemble the direct connection commands, as the direct connection version was the original.

14 This comment is true for most e-mail accessed information services.

15 The usual practice is for the server to be on a different machine from the client but this is not necessarily so – some host services point the client to a local server with local information first, and this server then has a pointer or pointers to distant servers.

16 NB: In both these cases the address includes the word 'gopher' – this is not always so.

17 BUBL – Bulletin Board for Libraries. A Gopher-based information service of interest to the library and information community (and others) running on the UKOLN machine at Bath. It was originally a 'traditional' Bulletin Board service, hence the title. BUBL is maintained mainly by voluntary effort (drawn from all over the UK via the network) and coordinated by Dennis Nicholson from the University of Strathclyde.

18 Some books claim this stands for *Very Easy Rodent Oriented Net-wide Index to Computerised Archives* and others that it is named after a cartoon character – maybe both are true.

19 NISS – National Information Services and Systems (previously National Information on Software and Services). A general information and gateway service of interest to the academic community (based at Bath on its own machine). Maintains a range of commercial and non-commercial databases, and provides links to other services in the UK, Europe and the US.

20 For example, the *Dow-Jones News Retrieval* service is available as a subscription WAIS service. Also, it should be noted that this problem of information sources being listed in databases but not available to casual users occurs in the Gopher and Web services as well.

21 All hypertext reading programs are known as 'browsers'.

22 In some cases a Web client will need to start external programs to handle certain formats, so what is possible with a specific client may depend on what other utility software you have available.

3

Gaining access

When thinking of Internet access there are three variables or dimensions to consider:

- the package of services required
- the form of service delivery
- the method of connection

The range of services available has been described briefly in Chapter 2. Although you might want e-mail alone, it is likely that a selection or package of these services would be more useful. Suggestions as to which service packages would be best suited to which users are given in Chapter 4.

As each of the other two variables or dimensions listed above has at least two options (and sometimes three or four), the permutations can get complex. Fortunately, some combinations are common and others are mutually exclusive for technical reasons, so the complexity is manageable with a little thought.

Forms of service delivery (or where the programs run)

We now need to separate the services you want to use from the way in which you access them, or to put it another way, the way in which they are delivered to you[1]. As an analogy, consider a restaurant: having looked at the items on the menu (the Internet services) we can now consider the way in which the meal is delivered. In one restaurant, your vegetable curry might arrive on three stainless steel dishes tastefully arranged on the table, and in another, it might arrive in three aluminium packages for you to take away. The food is the same but the form of delivery is different.

Unfortunately, while it is easy to distinguish between a restaurant meal and a take-away meal, it is not so easy to distinguish between the various forms of network service delivery. In every case, you use the service sitting at a keyboard looking at a screen. The crucial distinguishing factor is where the programs that you are interacting with (and that are providing the service you are using) are actually running. There are two main possibilities: either they run on a computer belonging to you or your organisation, or they run on the access provider's machine. The first option has two further subdivisions: the programs that provide the service either interact with the network (that is, remote machines) in real time or they interact in a batched form. This gives three[2] basic forms of delivery which are discussed below.

 Note: Although the 'batch' approach can be effective and very efficient it is limited to non-interactive service delivery and for this reason it is discussed last.

Terminal access (or log-on account or 'shell'[3] account)

In this form you have a log-on account on the access provider's machine and the service-providing programs all run on that machine. You interact with them via a terminal or, more likely, a personal computer running a terminal emulation program. If all you were running on the personal computer was a terminal emulator the difference between this form and the one described below in *Batch access* would be clear. However, some access providers supply the user with a package (known as an *off-line reader*) that eases access and blurs this difference. For example, the CityScape e-mail service is based entirely around such a package. These packages allow a user to download their e-mail and/or USENET News articles from the access provider's machine and then disconnect. They can then read the e-mail or News articles locally, compose any replies or new notes, and then re-connect and upload the new material. In addition, these programs usually manage the filing of e-mail, and other related files, on the user's machine. Some access providers offer their own tailored versions of such packages. Also, there are many third-party communications packages that offer similar features. There is no clear dividing line between these off-line reader services and true batch access. A clue is given by the underlying protocol – UUCP[4] indicates a batch access service, whereas the services based around proprietary off-line readers may use their own transfer protocol from the host to the local machine. Also, some companies offering a choice of off-line readers may also offer interactive services (telnet, FTP, and so on) via terminal access to their machine.

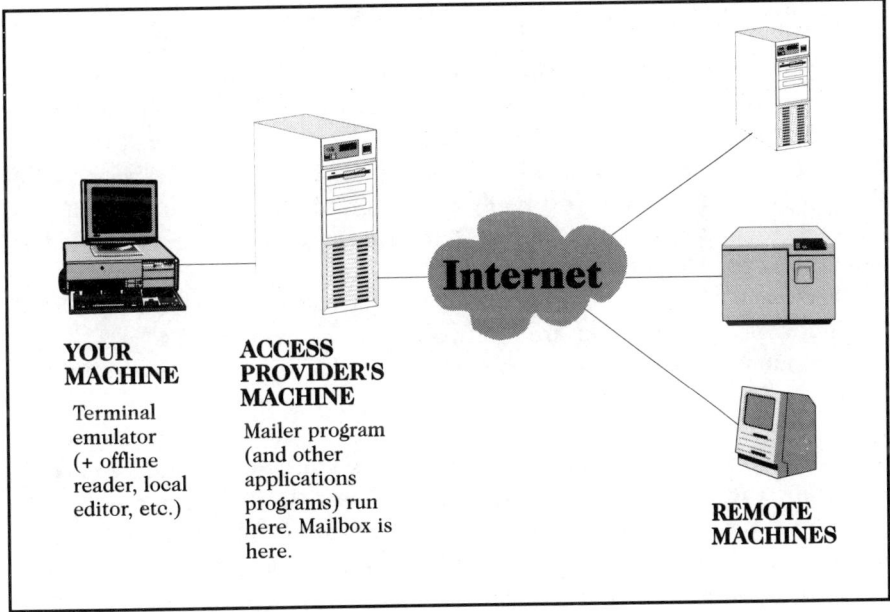

Figure 3.1   Illustration of the connectivity of terminal access

Although terminal access (without a good off-line reader) can be the most basic form of access it does have some advantages:

- You need no knowledge whatsoever of TCP/IP or the technicalities of Internet operation at the networking level. All this is dealt with by the access provider.

- If, for example, you have a PC at work, one at home and a portable, terminal access has the real advantage that you only have one e-mail 'in-tray' – the one on the host machine. If you have a service that provides e-mail from your machine you are in danger of having two or three parallel 'in-trays', one on each machine – not a good idea!

- When you are using FTP to fetch files from a distant machine you are not limited by the speed of your connection to the access provider's machine. You may still want to get the files onto your own machine, but if they contain text you can at least check its relevance before the final download. If the files are not already compressed, you should be able to compress them before you download to your machine, thus saving time and money (assuming you are using a form of connection like a telephone line or ISDN which is charged by time, or X.25 services that charge by kilobytes transferred).

- You have some inherent security since your machine is never directly connected to the network (though this does not stop you downloading programs or other files that may contain a virus!).

The major disadvantage of terminal access is that you usually have to work with the most basic interfaces to the services provided. E-mail, FTP and some of the more advanced services like Gopher work quite well this way (assuming you have at least VT100 emulation) but to get the best out of WAIS or World Wide Web you really need a graphics interface. Also, if you are used to a mouse-driven GUI[5] like Windows on the PC or that provided as standard by the Macintosh you may find a VT100 screen very unfriendly.

There is an exception to this if you have very high-speed access via a local area network, metropolitan area network, ISDN, or one of the latest high-speed modems, in which case you might be able to run X-Windows[6]. Expert opinion differs on whether this is a sensible option compared to the form of access described in the next section. However, if you have access to technology providing this degree of connectivity you probably do not need to be reading this guide! An example of X-Windows (in this case Xgopher) is shown in Figure 3.2 (yet another view of the BUBL Gopher server top-level menu as used in Figures 2.3, 2.4 and 2.10). As you can see, it is a full Windows 'point and click' interface.

Another disadvantage of terminal access, especially if you do not have a sophisticated communications package with a built-in editor and facilities for easy uploading, is that you may need to use the editor on the access provider's machine to enter the text for your e-mail. As most providers use UNIX machines, this may mean you have to use some of the most incomprehensible editors ever designed to make life difficult for the non-computer-specialist user!

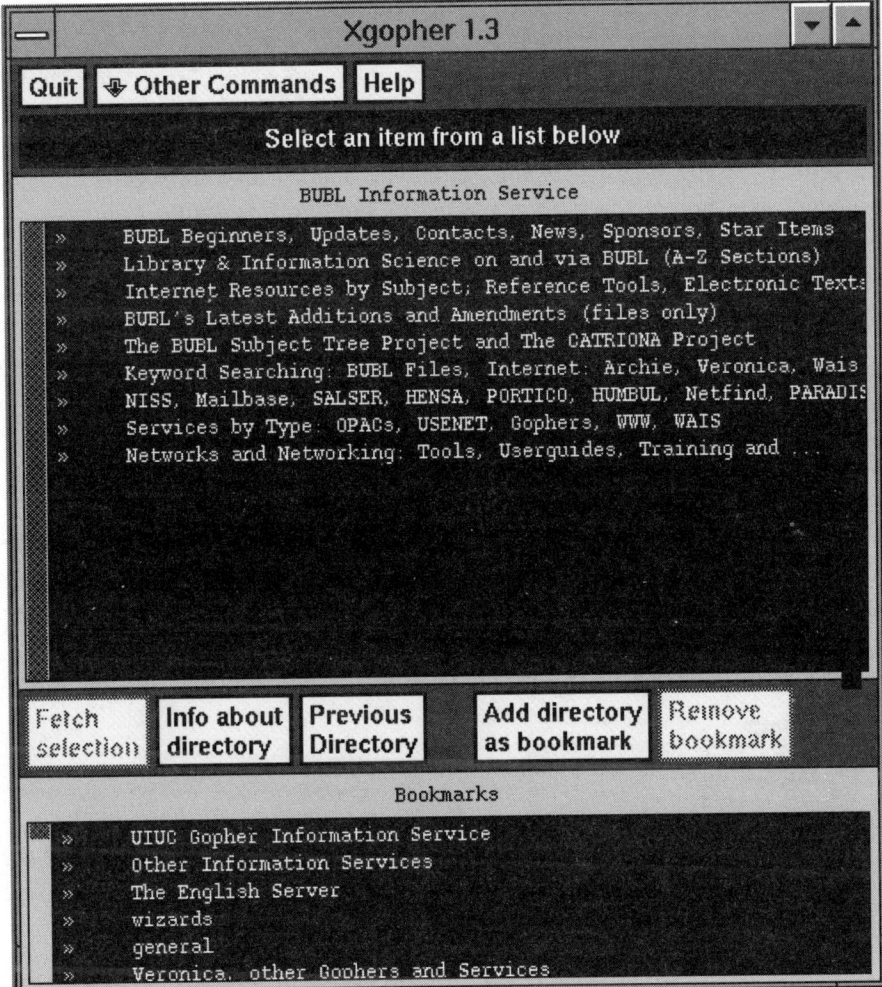

Figure 3.2   Example of X-Gopher on Windows screen

Full interactive (TCP/IP) access

For many network users this is the only 'real' form of Internet access[7]. Now the IP packets originate on, and are delivered to, your local (departmental or desktop) machine and you (or the organisation) are really *on the network* with a real IP address. This is a clean and simple model, with one exception – e-mail. E-mail machines are expected to be permanently connected to the network and running 24 hours a day. If you only have an intermittent form of connection or your machine is powered down overnight, distant machines are going to waste time trying to connect to your machine while it is unavailable – worse still, if your machine is unavailable for too long the sending machine will assume there is a real problem and return the mail to the sender, saying that it cannot be delivered. To get around this, access providers give you a temporary mail store on their machine (where you make your initial connection) which accepts incoming e-mail for you, stores it until the next time you connect to the network and then delivers it to you (it's like having a friendly neighbour who picks up your paper mail while you are away on holiday and then hands you the whole pile when you get back). The protocol that enables this is known as the Post Office Protocol (POP – not to be confused with Points of Presence). Access providers that offer this form of e-mail may mention POP or POPmail in the description of their e-mail service.

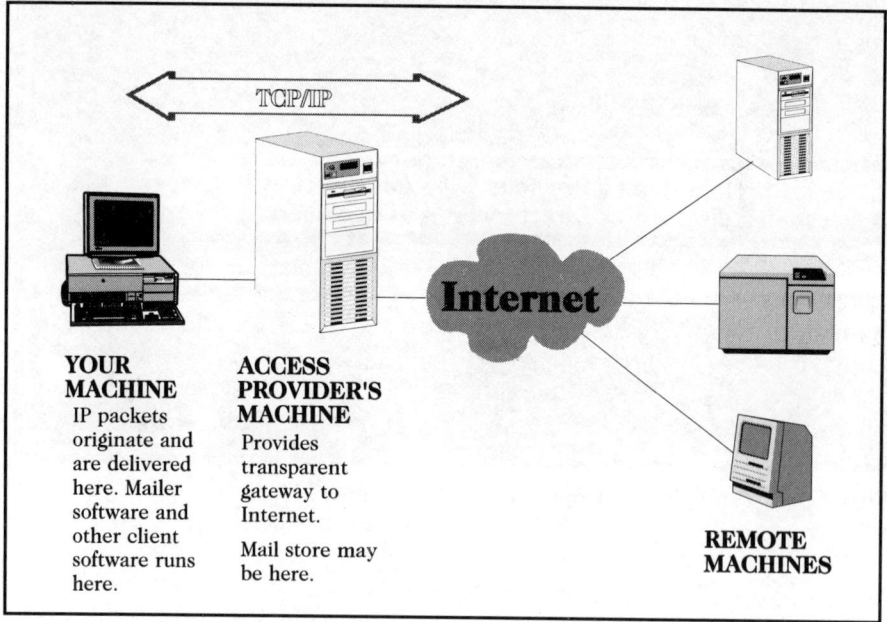

Figure 3.3   Illustration of the connectivity of I/P access

The great advantage of this form of service delivery is the total flexibility you have. You can run whichever mailer package you find easiest to use, you can pick and choose between the various Gopher, WAIS and other clients available for your machine, and so on.

The main disadvantage (especially if you are a small organisation or a lone user) is that you may have to set up and maintain all of this software yourself. Unless you are a systems person or have the right kind of support provided locally, this can be a little daunting. However, 'work straight out of the box' packages[8] are becoming available. There are two forms of IP addressing used with those packages – *static* and *dynamic*. With static addressing, either you get an address with the software package or it is assigned when you first connect to the access provider's machine – and from then on it is yours. Static addressing is always used with leased line type connections. With dynamic addressing you get (or may get) a different address each time you connect. This may seem odd but a little thought shows how it works. For running interactive sessions (telnet, Gopher, and so on) with another machine on the network, it only matters that your address remains constant for the duration of the session. However, some further thought will indicate that there would be a problem with e-mail if your apparent return address changed every time you logged in! This problem is solved by having your access provider's mail-store service as your e-mail address (as described above).

This approach of buying Internet access shrink-wrapped in a box is quite possibly the way things will develop in the future.

Batch (UUCP) access

Long before (at least by computing standards) the Internet became available to the general public there was (and still is) a worldwide e-mail and file transfer network known as UUCPnet. As its name suggests, it is based on the standard inter-computer UNIX file transfer protocol UUCP (UNIX to UNIX Copy Protocol). UUCP was designed to work over modems and noisy dial-up phone lines (though it happily runs over more sophisticated communications services, including the Internet, when it can). There are many areas of the world where the telecommunications services are not yet able to provide the infrastructure needed for networks like the Internet, and in these countries UUCP based networks are still common and extremely useful.

Despite its age UUCP works well and it is now available for non-UNIX machines like PCs and Macs running their usual operating systems. Some access providers offer e-mail and other services (usually USENET News) based on UUCP. The system works like this:

- There is a set of programs running on the local machine that enables the user (or more commonly users) to prepare e-mail and queue it for sending, receive incoming e-mail and read it, and/or access USENET Newsgroups.

- Periodically (anything from every few minutes to once a day or longer) the user's machine calls the service provider's machine and downloads any new mail or new USENET articles and uploads any outgoing mail or USENET postings. Another (usually more expensive) option is to arrange for the service provider to call the user's machine whenever there is any incoming mail – this is sometimes referred to as 'dial on demand'.

If the local system makes contact with its 'feed' reasonably often it is difficult for the casual e-mail user to distinguish it from a more sophisticated service using a permanent connection.

The great advantage of UUCP is its availability – all UNIX systems support it. In addition, it has been around so long that it is well understood and totally reliable. It also has a range of compression and other techniques that enable service providers to combine and compress files and deliver the whole package efficiently over a standard dial-up telephone line.

The major disadvantage of a UUCP based service is that it does not support true interactive services. This rules out the newer services like Gopher and WAIS (both these services can be accessed via e-mail – but you need to have the sort of patience required to play chess by post!).

Despite this, since UUCP and the basic e-mail software is available on all UNIX systems, any organisation with a UNIX system could consider UUCP based delivery as one quite cheap option for giving a group of users access to worldwide e-mail and USENET Newsgroups.

Methods of connection (or the initial link)

So far we have considered the range of network services offered and the forms in which they may be accessed or delivered. The final decision is how to connect to the access provider.

It may seem odd to leave the discussion of the initial connection to last but until you have at least a basic idea of the possible services and forms or models of service delivery it is difficult to explain easily the advantages of the various connection methods.

Looked at simply, there are only two types of connection, temporary (or *on demand*) and permanent. Unfortunately this simple dichotomy breaks down when one looks more closely at the options. While it is possible to argue that there is only one form of real permanent long-distance connection, the *leased line*, some temporary connections are so reliable and can make and drop the link so quickly that for many purposes they are indistinguishable from leased lines in operation.

The methods of connection will be covered in approximate order of cost and bandwidth supplied.

Bandwidth (another technical idea explained)

As bandwidth requirements are important when considering the initial link, bandwidth needs to be defined here for those not entirely sure what it means. The bandwidth of any link refers to the capacity of that link to carry data. The usual analogy is to water-pipes: wide water-pipes can move more water than narrow ones in the same time. In the same way, wide bandwidth links can move or carry more data than narrow bandwidth links in the same time. Capacity is measured in 'bits per second'[9] (or the usual multiples thereof, kilobits, megabits, and so on), but the 'per second' is often dropped as it is assumed to be understood. It also needs to be made clear that what is being referred to is *capacity*, and not, necessarily, the actual use. A megabit capacity link may in operation not carry more than 250 kilobits (for example) during any one second but it has the capacity to carry more if needed. Finally, the terms: wide bandwidth, high bandwidth, high speed, high capacity (likewise, low bandwidth, low speed, and so on) are used interchangeably and can be considered to mean the same.

A single e-mail user could probably make do with a basic 2400 baud connection as long as he or she didn't need to download large text files very often. To use something like a Mosaic Web browser the same user would need at least a basic 9600 baud connection, preferably with some sort of compression (discussed below). Multiple e-mail users on a shared machine or a LAN could work with 9600 baud, but for many parallel interactive sessions one would need to think of providing bandwidth in the tens or hundreds of kilobits, or even megabits.

A graphical illustration of bandwidth, portrayed as pipes whose diameters indicate a range of bandwidths commonly available, is given in Figure 3.4. To be exact the next major step above 64K is 2Mbits, not 1Mbit, but this would have meant that the 2400 baud pipe would have been too small to draw as a 3-D pipe and would have become just a line!

So far we have considered basic bandwidth. This can be improved upon by using some form of data compression technique on the data as it is transmitted. For modems this usually means one of the two standards, MNP5 or V.42bis (both are often available on the same modem). Using these techniques the notional bandwidth for a link can be improved by factors between 2 (MNP5) and 4 (V.42bis). However, due to the underlying mathematics of these techniques, this increase is only true for textual data which has a high degree of redundancy. Binary data or text files that have already been compressed before transmission will not show this degree of improvement.

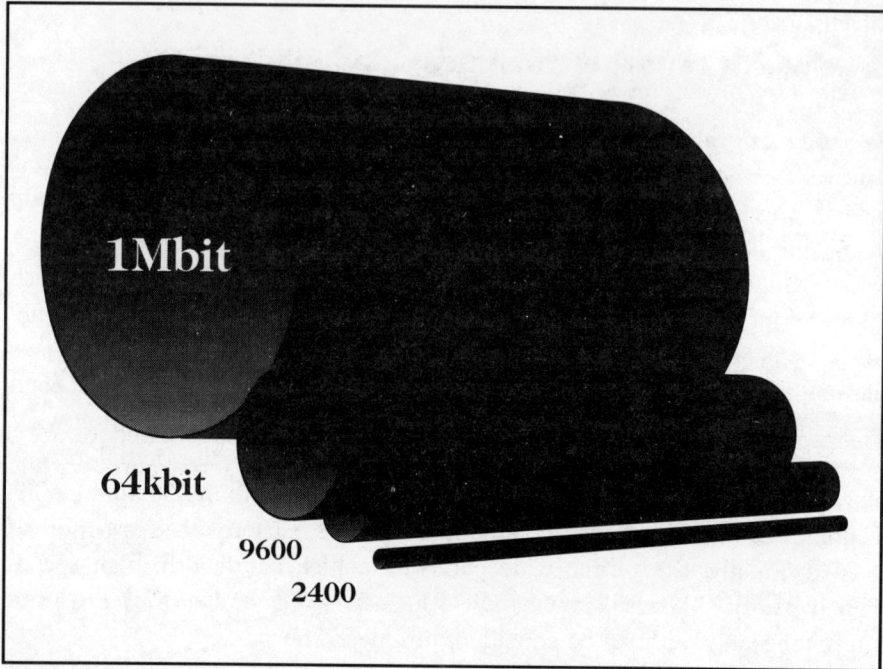

1Mbit

64kbit

9600

2400

Figure 3.4   Graphical illustration of bandwidth as pipes

Dial-up or on-demand links

With this form of connection you do not have a permanent connection to the network; you connect to it (or to your access provider) as you need to. There are three ways of doing this, *modem*, *ISDN*, and *X.25*.

Modem links

Most readers are probably familiar with modems. Those who need a basic introduction should read Appendix C. Dial-up via modem is the way most single users and small companies will connect to the access provider's machine and thence into the Internet.

The advantages of the modem are:

- It is relatively cheap to start (high-speed modems [9600 baud and above] start at less than £200 including VAT).

- It is easy to set up and use.

- The underlying communications technology is widely available (ordinary telephone lines).

The disadvantages are:

- The maximum bandwidth available is relatively low (currently 14400[10] baud without data compression and many services do not support above 9600).

- You pay by time used (exactly as with a voice telephone call), in addition to any fixed and connection charges your access provider may charge.

- You cannot guarantee connection, as the access provider has fewer modems than customers. The number of modems provided is calculated from the service level (service availability indicated in the contract) and a standard pattern of use.

The problem concerning maximum bandwidth may be exacerbated for users of older PCs combined with external modems. Their serial ports usually contain the 8250 UART chip which does not run very reliably above 9600 baud. Newer machines should have a 16550 UART. If the speed of the serial port is the limiting factor it is now possible to buy plug-in serial cards fitted with the later UART. This will be true for all methods of connection that use the serial port as the point of connection to the PC.

The problem outlined in the third item of the list above can be avoided by having a telephone number and modem dedicated to your use at the access provider's site or local POP (Point of Presence – explained in Chapter 5). This is sometimes called a *dedicated line* or *reserved line* and should not be confused with a *leased line* as described below[11].

The above paragraphs assume you are calling the local POP of the access provider direct – there are other variations. For example, the access provider may have X.25 access. Even if you do not have direct X.25 access you may have (or could buy) access to one of the dial-up X.25 services like BT's DialPlus or the equivalent in the UK from Mercury, AT&T or Sprint. There is a similar service from Telecom Eireann in Ireland, known as Eirpac. Assuming you do not live within a local telephone call of a possible access provider, you could use one of these services at least to limit your telephone bill to a local call from most of the UK or Ireland. Access via these services may still be limited to 2400 baud in some areas in the UK, but the majority of nodes now offer 9600 baud plus V.42bis or MNP5 data compression. GNS DialPlus will also offer 14.4 Kbaud sometime in 1995. It should be remembered that you will get an additional bill from the dial-up X.25 service supplier related to the amount of data transferred, but the combined bill could work out at less than the cost of a long-distance telephone call. However, because of the speed limitations set by some dial-up X.25 services there may be situations where this is not true. As a rough guide – if you expect to spend much time actually connected to the access provider's machine and you do not have local call access to one of the access provider's POPs, it is worth checking out the possible use of a dial-up X.25 service.

Remember: the modem link is just a carrier – what you run over it is up to you. You can use it for simple terminal access or you could run one of the protocols (SLIP or PPP[12]) that enable you to run IP over a serial link (which is what a modem provides).

ISDN (Integrated Services Digital Network) access

ISDN is, as yet, an underused service from BT. It can be delivered over the existing telephone wires and provides a true digital network into the average office or home. The basic ISDN connection replaces your existing analogue telephone connection – instead, you have an ISDN box to which you can connect true ISDN equipment (including telephones, high-speed faxes and ISDN cards for PCs) and in addition, if the connecting adaptor allows for it, analogue equipment (ordinary telephones, and so on). The basic ISDN service (ISDN 2) provides two 64 Kbit/sec data channels and a 16 Kbit/sec control channel. The two channels operate in parallel, so you could send data down one while having a telephone conversation over the other. The costs of ISDN calls are identical to the equivalent telephone call (taking into account time of day, length of call and distance)[13]. For companies with a greater bandwidth need there is the ISDN 30 service with thirty 64 Kbit/sec channels. This is obviously much more expensive in terms of hardware requirements and rental.

The two major advantages of ISDN are the bandwidth of the link and the speed with which connections can be made and dropped. While a modem can take tens of seconds to set up a call, an ISDN call can be set up in less than a second. This means you can have applications running on a local machine that need access to certain files on a distant machine quickly (within a few seconds), and cannot continue with their operation until they have this access, without the need for a permanent connection. Obviously, as ISDN calls are charged by time, there comes a cross-over point, if you (or your application) need to access the distant system very often or to hold the connection for a long time, when it becomes cheaper to move to a leased line.

The major disadvantage of ISDN is the cost of installation (around £400) and line rental (currently around £84/qtr compared with £20.16/qtr for a standard residential telephone line[14]). To this you may need to add the cost of any additional ISDN related equipment (for example, adaptors to connect existing analogue hardware), which is still more expensive than comparable analogue equipment. These costs may fall as ISDN becomes more common.

As with modem links, ISDN is just the carrier. You could use it just for simple terminal type access, but with such bandwidth available interactive IP or X-Windows terminal access from the local machine would seem a more sensible use.

X.25 access[15]

If your organisation already has access to X.25 network services, many Internet access providers allow you to make an X.25 call direct to their services. In some cases you only have a form of terminal access or file transfer service, but others offer IP over X.25. If this is available, it should be indicated in the Internet access provider information pages in Part Two.

Permanent network connection

Leased line

If you want to be a permanent part of the Internet, or your use of on-demand type connections is so great that the costs are equal to or greater than the rental of a leased line, this is the way to go. A leased line is a dedicated circuit provided by one of the communications companies (BT, Telecom Eireann, Mercury, Sprint, and so on) with a specified bandwidth availability. Leased lines are charged by the bandwidth provided and the length of the connection between you and the access provider's nearest POP.

The decision to use a leased line rather than an on-demand form of access will be based on both cost and service availability. If you have a LAN with ten users, with at least one of them needing interactive access to a network resource at any time, then leased line access makes sense – if you have one or two users with only occasional need for access then it certainly does not.

Although, like the foregoing forms of connection, the leased line is just a carrier and you could run a range of communications services over it, you will almost certainly run native IP back to your access provider who will link you seamlessly into the Internet. When looking at this level of service there are many costs in addition to the rental of the line. You will be running a LAN or large multi-user machine, and the leased line will need to be terminated at both ends with appropriate hardware. The access provider will obviously not charge you the capital cost of the equipment at their end, but the outlay will be reflected in the fees charged and the form of contract required. In return, the access provider will often be prepared to rent the hardware needed at your end to you as part of the package (with an option to buy after a fixed time). Further, the installation of a leased line is non-trivial, especially if you have a LAN and you want all the machines on the LAN to have their own IP addresses. In this instance, unless you have considerable in-house expertise, this is a case for consulting the access providers and/or one of the new breed of Internet access consultants[16], to analyse your needs and do a full-scale design before decisions are made. The main problem with leased line and native IP access is that of security. With this level of connectivity your system is totally visible to the whole network. Although the designers of operating systems have become more security conscious over the past few years, it is recommended that if you intend to have a full TCP/IP connection then you need to consider taking precautions against illicit access. This is a complex technical matter outside the scope of this small volume – take expert advice!

One use for leased lines in the past was when you were providing a service, for example, database, booking service, and so on, which you wanted others to be able to access from the Internet. However, if access is likely to be light, other services such as ISDN (with the access provider initiating the call) might be an option. The incoming service user would be unlikely to be aware that an on-demand service was being used, due to the speed of making the connection.

Notes

1 You can think of a service being either accessed by you or delivered by the access provider – it is 'two sides of the same coin'.

2 As is usually the case in computing nothing is ever this simple – there are always many ways of achieving the same end. However, these three forms encapsulate sufficiently different models of delivery to be useful when thinking about the services offered.

3 The term 'shell' refers to a *user-interface* or *front-end* program with which you interact when you log in to a shared machine. This may enhance and/or hide the underlying operating system (usually UNIX).

4 UNIX-to-UNIX Copy Protocol.

5 Graphical User Interfaces.

6 X-Windows is another client/server animal. Part of it runs on your machine and looks after the screen and human-machine interface and the other part runs on the distant machine looking after the interaction between the application program, for example, a Web client and your machine.

7 For such purists, the form of access described in the previous section is just about acceptable as you are running TCP/IP from the access provider's machine, but that described in the following section is not true Internet access at all as you are not running TCP/IP.

8 O'Reilly launched their *Internet in a Box* package at the *Internet World* Conference in New York, December 1993. Latest information on shipping for the PC version is third quarter 1994. CityScape have recently (July 1994) launched their IP.GOLD service which delivers an e-mail package, a News reader, and a Web browser on a single disk.

9 You will often see the word *baud* used as if it were equivalent to *bits per second*. Strictly speaking, this is only true for binary (or two-state) communication systems, but for most purposes this blurring of the two terms is acceptable.

10 28800 baud is becoming available but this is still uncommon. Also, it must always be remembered that both ends need to support the speed.

11 However, *dedicated line* is often used to mean *leased line*. *Leased line* is the preferred term, as it is unambiguous.

12 SLIP = Serial Line IP and PPP = Point-to-Point Protocol.

13 To be exact – each 64Kbit/sec channel is charged at this rate (since each can carry a separate call).

14 However, you do get two channels for this.

15 It is not clear if X.25 access should be here under *dial-up* or *on-demand links* or below under *permanent links,* as X.25 services are themselves delivered over leased lines. As the actual X.25 calls are made and broken they are considered as temporary connections.

16 There are two companies listed in Part Two that call themselves Internet access consultants first and access providers second, ElectricMail Ltd and Motiv Systems Ltd.

4

Making the right connection

In this chapter, some typical user needs are considered along with the options available to them, and the most sensible solutions are outlined.

Obviously the number of variables is large, – number of users, form of access required, services required, degree of expertise available, and so on, and hence the possible number of permutations is enormous. For practical reasons are only a small subset (three) will be considered, though the options will be outlined and discussed in each case. Although the number is small it is felt that the following scenarios give some insight into the problems/solutions of other possible users.

Single user – home access

It is assumed that any home user is using a PC, Mac or other machine of equivalent power and has limited technical knowledge.

The main options are between the services required and the form of service delivery. The method of access will almost certainly be by modem. ISDN is possible but unlikely to be cost effective (at current rates and taking into account the additional hardware required) unless there is a real need for the bandwidth.

If e-mail only is required, a simple terminal access type of service is probably the best bet – especially if the access provider can provide, or recommend, an off-line reader to minimise the time spent on-line.

If more extensive access is required, including extended use of network interactive services like FTP and Gopher, the choice depends on the expertise of the user (how comfortable he or she is in setting up IP on their machine). Although it is perfectly possible to use tools like Gopher and Web via a VT100 interface on the access provider's machine, the PC or Mac based clients are undoubtedly more user-friendly. Therefore, the final decision may be made on the need for or desirability of a graphics interface. With products like O'Reilly's *Internet in a Box* and CityScape's *IP-Gold* service becoming available, the choice may tilt more towards IP from the user's machine – at present it is probably still more for those with some technical knowledge (or an obliging friend!).

Something to consider: when using a tool like Gopher the time spent searching (and hence on-line over the telephone line) is approximately the same whether you are running the client on your machine or the access provider's machine. However, with the latter you may still need to download the result to your machine – an extra chore taking more time[1]. In addition, the graphical interface of a PC or Mac Gopher should be easier to use and possibly more productive.

Small group – E-mail only access

For a small organisation, with a few users requiring e-mail only access, the decision will probably turn on whether there is a LAN. If there is, it would make sense to go for a mail machine on the LAN which gives access to the Internet via some form of on-demand connection. If there is an existing LAN based e-mail service, the possibility of using this as a front-end to the Internet mail connection should be given serious consideration. There are companies (at least two are listed in Part Two) who provide the software and expertise to make this transparent to the end user. If you can go from local e-mail to worldwide e-mail without needing to retrain any of the users, this has to be a 'good thing'. It is possible that a service based on a UUCP connection combined with a high-speed modem would be perfectly adequate and cheaper than some of the other options.

If a LAN is available, but no local e-mail is currently used or needed, then another option would be to have a networked modem server which would allow each user to connect to a terminal access e-mail service. This is not ideal as there can only be one user at a time, but it does share both the modem and the telephone line.

If a LAN is not available, it might be worth considering one if all of the team are going to need regular access to e-mail, thus enabling the options outlined above – if there are more than five or six users this would probably be cheaper than giving all the users a modem on their machine, even if telephone lines are shared using a PABX.

If the LAN option is simply not possible because the users are distributed across various sites, individual terminal access to a simple e-mail service via a modem is probably the best option.

Larger group – LAN and full access required

We are thinking here about 20+ users on one or more sites with a fully operational LAN (with inter-site connections) and internal e-mail. All the users need or would benefit from full Internet access, including Gopher and Web connections.

The method of connection is defined by the demand – it must be ISDN at least and probably leased line.

The major decision is whether to offer IP to each user machine or to have a central machine to which the others have basic terminal or X-Windows access. If the current e-mail service is based on such a model, this might be very acceptable to the users. If the current internal e-mail system is distributed across all the machines, the users might find accessing a central system unduly restrictive and the actual interface not up to the standard they are used to. However, the X-Windows approach can deliver a top quality graphic interface to compete with any Windows based package.

The final decision will probably depend on many factors. With IP to every machine security would be a major consideration, while if it (IP) stopped at the central server, it would be more manageable. Also, the LAN would need to be either an IP network itself or able to handle an IP network running over it.

With all the options and the costs involved this is definitely a case where, unless there is plenty of in-house expertise, external advice should be sought – and happily paid for!

Notes

1 However, there is a counter-argument to this outlined in the section entitled *Terminal access* in Chapter 3.

Part 2

Internet access providers in the UK and Republic of Ireland

Internet access providers in the
UK and Republic of Ireland

5

Internet access providers in the UK and Republic of Ireland

Introduction

Information on each of the known UK IAPs is listed below in a standard form. It was gathered from various on-line and paper sources and from the literature (where available) of the companies concerned.

This list is not exhaustive but is as complete as possible. There are other services accessible in the UK, mainly from US based companies, but in these cases the support may be minimal as the companies concerned see the UK and Ireland as just one small part of a global market. All the companies listed have staff and support (where offered) in the UK or Ireland.

Where possible this information has been checked with the companies concerned, but in such a fast-moving area it will no doubt date quite quickly.

Missing major players

Until quite recently the large communications companies operating in the UK and the Republic of Ireland (British Telecom, Telecom Eireann, Mercury Communications, Sprint International, AT&T, and so on) have watched the growth of the Internet and provided the underlying infrastructure but not offered Internet services themselves. This is changing. BT has already announced that it will be providing some Internet related services from late 1994. France Telecom is providing IP services in France and may move to offer similar services in other parts of Europe, including the UK.

Attempts have been made to contact these companies to ascertain their intentions but in most cases they are still in the planning stage and they were unable to supply details. However, it is clear that many of these companies are seriously considering offering IP access, and possibly other Internet services, soon. In most cases these will be aimed mainly at the corporate market. To enable readers to get the latest information in this important growth area, there is a list of contact names and numbers (where known) in Appendix B.

 STOP PRESS – Microsoft has recently announced that it intends to become a network access provider. The proposed service will be similar to CompuServe with shopping, news and financial services, plus reference works and an online encyclopedia, in addition to Internet access. The service will be tied-in with the launch of Windows 95, the latest version of Microsoft Windows. As yet it is not clear if this service will be available in the UK.

Explanation of the Companies Information Pages

Information collected on each IAP has been translated into a standard form. This section describes the possible contents of each field in the form.

The layout of this section is based directly on the layout of the form.

Name

Each record starts with the trading name of the company or service. This name is used to order the list alphabetically. While in most cases this is unambiguous, sometimes there are problems. For example, the

PC User Group was originally known as the *IBM PC User Group* and their e-mail address still has 'ibmpcug' in it. Further, the *PC User Group* now offers two distinct (but overlapping) services (CONNECT and WINnet) which are sold separately. In situations like this, the original service is listed under the trading name of the company (PC Usergroup – CONNECT), but the separately advertised and sold service (WINnet) is listed under its name. In the case of CityScape, which has recently launched a new IP service parallelling its original e-mail service, the new service is listed separately but includes CityScape as part of its name, so the two services are listed consecutively.

Address

The trading address of the company. In some cases this is not known as some companies trade almost entirely electronically – including on-line joining and billing to a credit card.

POPs

Points of Presence – access points to the service. Many access providers can only offer local call rates to users geographically close to their site. The larger companies are installing POPs in major cities or population centres around the UK. This means a user calls the nearest POP to give the lowest call charges. Having a POP as close as possible also affects leased line users as the cost of these lines partly depends on the distance from the user to the POP.

Although most companies list POPs according to the city where they are situated, some list the places from which users can obtain local call rates – this can be confusing. This is indicated where it is known to be the case.

Another option is for the company to offer X.25 access, which means they can be accessed via one of the X.25 services (BT GNS, Mercury 5000, and so on) direct, or via BT DialPlus or other dial-up X.25 service. This is covered more fully in Methods of Connection below and in Chapter 3.

Proposed POPs (by end 1994)

POPs listed under '*POPs*' are available now. In order to extend the usefulness of the Guide each company was asked to supply information on POPs it intends to provide in the near future. Unfortunately, few replied but those who did are listed.

Basic services

The basic services commonly available have been explained in Chapter 2. The company concerned may not see these as *their* basic services but they are basic from an average user's point of view. For example, *On-line Entertainment Ltd* are currently primarily an entertainment company, but they do also provide an Internet access service which is felt to be a basic service from the point of view of the general user.

Additional services

These may be various, and it is sometimes difficult to decide whether a service is *basic* or *additional*. In most cases, the decision is based partly on the apparent importance given to it in the company's own literature, and partly on the approach taken in the description of these services in Chapter 2 where e-mail, FTP and telnet are described as 'basic services'.

Definitions of those services not covered in Chapter 2 are included below (though there will obviously be some overlap):

Alert service
Some companies offer the facility to alert users to new incoming e-mail via fax, telephone or pager.

Archie
The host machine has an Archie client.

Batch FTP
This is a service whereby requests for files to be retrieved from other sites are queued locally and the files are collected later as part of a batch operation. The files retrieved are then placed in the user's *FTP space.*

Dial-on-demand
Where a service is delivered to the user's machine (for example, mail or USENET News), and the user does not have a permanent connection, some companies arrange for the host to dial the user's machine either as incoming items arrive or at preset times.

E-mail access to FTP archives
This is the ability to access remote FTP archives by sending e-mail containing FTP-like commands to a local or distant service. It is offered by some companies who do not offer full FTP access (or not as part of a specific service).

Fax gateway (or outgoing Fax)

Some services allow users to send a fax as if it were e-mail, which can be particularly useful if you are sending the same fax to a long list of numbers. Other services go further and send the image file over their own network to the node nearest to the destination of the fax, before passing it to the standard telephone lines. For international calls this should reduce cost and may improve the quality of the delivered image. However, recipients cannot reply via the same path. Because Fax is an image based service there is no simple, standard way to indicate the target e-mail address in computer-readable form in the incoming fax.

 STOP PRESS – The London Host have recently announced a service that does allow an incoming fax to be delivered to your e-mail box in image form. See their page in the Internet access providers list for contact details.

FTP space

This is an area of disk space on the host machine where users of services that provide FTP from the host can 'park' and inspect files before downloading them to the user's own machine. This facility may also be made available in services that have FTP both from the user's machine and the host machine, thus enabling FTPing of files across the network at speeds in excess of the speed of the link between the user's machine and the host. These files can then be inspected (and possibly compressed if this has not already been done) before transferring to the user's machine.

Gopher

Host has an Internet Gopher client service.

Host services

Some access providers can put material from clients on their network information servers, for example, FTP, Gopher, WAIS, World Wide Web, and so on. So, for example, a company could have its advertising literature or price list up on a Web server, and so visible to the whole network, without needing to have its own machine or any expertise in this area.

Indirect FTP

This is where the service does not support FTP from the user's machine but does have FTP from the host. *See* also *FTP space*.

IRC or irc

Host supports Internet Relay Chat. This is a form of interactive real-time conferencing. It's a little like a cross between e-mail, or BB based conferences, and CB Radio – you join an ongoing discussion about a theme (or start one). The comments from each contributor appear on your screen, and your comments on theirs, as soon as they are sent.

Local Bulletin Boards (or local Conferences)
These are BBs on the host where users can exchange messages or 'discuss' specific topics. They are not visible off the host.

Local software archive
Some companies offer archives of *freeware* or *shareware* software on their system. The exact delivery mechanism to the user's machine will depend on the service being used. This software may be provided by the company from its own resources, bought in on CD-ROM or collected from network archives (or some combination of all three). NB: This software will probably have been checked for viruses but unless this is guaranteed, it should be checked with a suitable virus-checker program before use.

Menu front-end
Host has the option of a menu-based user interface to shield the nervous from UNIX.

Off-line reader
This is a package available from the company that allows the user to download mail or News from the host and read it on the user's machine. Most useful where the company charges for connect time or the user has to make a long-distance call to access the nearest *POP*.

On-line readers
This usually refers to USENET News reader packages. Different readers offer different 'views' and functionality. It may also be used to refer to a variety of e-mail readers.

Public FTP archive
Host supports anonymous FTP and is able to make users' documents available in this manner if required.

Sub-net or sub-domain addresses
Where the user is an organisation with a LAN or WAN of its own, the larger companies offer a service whereby they arrange for the provision of IP addresses for the machines on the user's network.

Telnet login
Host supports incoming telnet access from the Internet.

USENET News
Service gives access to the USENET Newsgroups. These may be read on the host (*on-line reader*) or on the user's machine (*off-line reader*).

UUCP or NNTP

Mail and USENET News can be delivered to the user's machine using the UUCP Protocol (*see* Chapter 3). NNTP offers a similar service for USENET News but enables the receiving machine to have more control over what is sent by interacting with the sending machine during the transfer.

Telex

Users can send and receive telexes on the service.

Users own mailing lists

Some companies allow users to set up their own private or public mailing lists so they can arrange their own discussion groups.

WAIS

Host has a WAIS client service.

WWW

Host has a World Wide Web client service.

Help/Support

This will indicate the forms of help/support available (telephone, e-mail, and so on) and the hours it is accessible. If there are special services for newcomers, this will be mentioned.

 If this space is blank it does not necessarily mean that the service does not provide help/support, merely that it was not mentioned in the literature or other information sources consulted during the compilation of this list.

Method of connection

As discussed in Chapter 3 – the methods of connection offered will be listed. If any methods incur extra cost that will also be indicated here.

Some of the companies that support X.25 access have their own accounts on X.25 dial-up services like BT GNS/Tymnet or Mercury 5000 and allow customers to access them using this account. This may provide local call rates to the X.25 dial-up service POP but there will be an extra charge from the Internet access provider for this service.

Form of service delivery

As discussed in Chapter 3 – the form or forms of service delivery will be listed.

Documentation

If the amount or form of documentation provided for users is known, it will be listed here.

Minimum hardware/software requirements

Most services can be accessed by any computer with a standard communications package and a modem. Some services require proprietary software (usually supplied as part of the service).

Fees

The latest known fees are indicated. In many cases, the services listed will be a selection of the possible options, due to limitations of space. Some companies offer a small set of central services plus a set of options and the exact fees will depend on the options chosen. In addition to the monthly, quarterly or annual fees, many companies have an initial set-up fee.

The forms of charging model vary widely from pure connect time charging, through connect time plus flat fee, to flat fee with no connect time or volume charges (and other permutations).

For more information

The following pages are only summaries; for detailed information on a specific service you can contact the company in various ways. *Tel, Fax* and *E-mail* are self-explanatory.

As the majority of the readers of this Guide are likely to be in the UK, all telephone and fax numbers are given from a UK viewpoint – UK numbers are in UK internal form and Republic of Ireland numbers are in their international form. This should avoid confusion when city or area codes are the same; for example, '021' refers to Birmingham in the UK and Cork in the Republic of Ireland.

The other options in this field may need explaining:

FTP

As described in Chapter 2, it is possible to set up anonymous FTP services containing publicly accessible files. Some of the companies listed use this facility to make detailed information concerning their services available.

Dial-up

Some companies offer information on-line which can be accessed by dialling in to their service using a standard communications package and modem and logging in with a guest id or picking a suitable option from the initial menu. If a guest id is needed it is indicated. Many of the companies also offer the facility to join in this way. This must be the ultimate in self-service – using the service to join the service with no human intervention at all!

Web or WWW

As with FTP above, the company has a Web server with information about available services.

Comments

General comments on the company, any changes of name, recent or forthcoming changes to services, and so on.

Also, if there is a need to refer to other records within this list, this is included here; this may happen when two services are related, say by being provided by the same company but sold as separate services by that company. This is also used to indicate that one company's product is an enhanced version of another's, or based on the underlying services provided by another.

Network access providers list

ALMAC BBS

Address:	ALMAC BBS Ltd, Heathpark, 141 Bo'ness Road, Grangemouth, FK3 9BS Scotland
POPs:	Grangemouth
Proposed POPs (by end 1994):	
Basic services:	E-mail, FTP and telnet (from host machine)
Additional services:	USENET News, access to other e-mail networks including FidoNET, off-line readers, extendedUSENET service includes ClariNet (the on-line newspaper)
Help/Support:	
Method of connection:	Dial-up, ISDN
Form of service delivery:	Terminal access
Documentation:	User Guide
Minimum hardware/ software requirements:	Any computer that can run a basic communications and terminal emulator package + modem
Fees:	Silver Subscription (Internet e-mail and USENET + 1 hour per day on-line included) £45 p.a.
	Gold Subscription (as Silver + ClariNet + 2 hours per day on-line included) £75 p.a.
	(Prices include VAT)
For more information:	Tel: 0324 666336
	Fax: 0324 665155 Dial-up: 0324 665371
Comments:	

BBC Networking Club

Address:	BBC Networking Club, PO Box 7, Broadcasting Support Services, London W3 6XY
POPs:	London, Cambridge, Edinburgh, Manchester, Bristol and Birmingham
Proposed POPs (by end 1994):	
Basic services:	Initially a BB service, expandable to full TCP/IP access (telnet, FTP from your machine)
Additional services:	The BB service will include information on BBC broadcast programmes and educational information. Archive of software packages proposed.
Help/Support:	
Method of connection:	Dial-up
Form of service delivery:	Initially terminal access to BB, moving to full IP service from own machine soon
Documentation:	
Minimum hardware/ software requirements:	Proprietary 'starter kit' communications package available for PC, Mac and Acorn Archimedes + modem. Starter package for Amiga in production.
Fees:	PC or Mac starter kit £25 + £12/month Archimedes starter kit £35+£12/month
	For current Internet users with telnet access to the BB (known as Auntie) will be £5/month (+VAT)
For more information:	*Tel:* 081 576 7799
	Fax: 081 576 9666 (information)
	Fax: 081 993 6281 (to subscribe)
	E-mail: emma@bbcnc.org.uk
Comments:	Launched on the BBC2 programme The Net. Initially a bulletin board with information about BBC broadcast programming and activities of the Club. Subscribers will either get further software as part of the starter kit or download it from the BB to provide a full IP service from their own machine.

British Telecom – Public IP Service (Pre-launch name)

Address:	BT Managed Network Services, Network House, Brindley Way, Apsley, Hemel Hempstead, HP3 9RR
POPs:	London and Manchester (initially – others planned)
Proposed POPs (by end 1994):	
Basic services:	IP to your machine
Additional services:	E-mail feed, News feed, Telnet, FTP. On-site configuration and network set-up
Help/Support:	Primary support – 9 a.m. - 6 p.m. Secondary support (fault reporting) 24 hours
Method of connection:	Dial-up (dedicated access port), ISDN, leased line, X.25, Frame-relay, SMDS
Form of service delivery:	PPP over dial-up, TCP/IP over leased line, IP over Frame-relay, IP over SMDS, IP over ISDN, IP over X.25
Documentation:	
Minimum hardware/ software requirements:	Any machine capable of running PPP plus modem, or hardware based router connection
Fees:	KiloStream – Connection £999, £5000/year
	Frame Relay Gateway – Connection £999, £3500/year
	SMDS Gateway – Connection £999, £3500/year
	X.25 Gateway – Connection £999, £2000/year
	ISDN – Connection £499, £3000/year
	Premium Dial-up – Connection £499, £1750/year (plus any communications charges, leased line.)
	(+VAT)
For more information:	*Fax:* 0442 237353 (Att. Tony Sweet) *E-mail:* info@bt.net or sales@bt.net

Comments:	Service launched November 1994. The initial service is designed for high- bandwidth access to the Internet and aimed at companies who will make substantial use of the service. A low-cost dial-up service may be introduced later.

CityScape – E-mail Service

Address:	CityScape Internet Services, 59 Wycliffe Road, Cambridge, CB1 3JE
POPs:	London, Cambridge, Edinburgh, Manchester, Bristol and Birmingham
Proposed POPs (by end 1994):	
Basic services:	E-mail using proprietary off-line reader
Additional services:	E-mail access to FTP archives, local Windows software archive, provision for users own mailing lists, and so on.
Help/Support:	24-hour support for all technical and e-mail problems from CityScape, plus 24-hour support for POPs and Internet connections from PIPEX)
Method of connection:	Dial-up
Form of service delivery:	Proprietary communications and off-line e-mail package
Documentation:	
Minimum hardware/ software requirements:	PC 386/486, Windows 3.1 and Hayes compatible modem (9600 baud recommended)
Fees:	£400 per annum or £150 for 1st qtr then £100/qtr thereafter No on-line or other charges (+VAT)
For more information:	*Tel:* 0223 566950 *Fax:* 0223 566951 *E-mail:* sales@cityscape.co.uk

Comments: Based on a customised version of the 'Mail-It' package from Unipalm (the parent company of PIPEX). This has automatic 'new mail' checking, binary file transfer (to allow exchange of word-processing and similar files), and so on. See also CityScape IP Gold service.

CityScape – IP.GOLD Service

Address: CityScape Internet Services, 59 Wycliffe Road, Cambridge, CB1 3JE

POPs: London, Cambridge, Edinburgh, Manchester, Bristol and Birmingham

Proposed POPs
(by end 1994):

Basic services: IP (to your machine)

Additional services: Includes software for e-mail, telnet, USENET News access, World Wide Web browser (also able to access Gopher, WAIS and FTP based services), and so on.

Help/Support: 9 a.m. - 6 p.m. voice support from CityScape, 24–hour e-mail based support for POPs and Internet connection from PIPEX

Method of connection: Dial-up

Form of service delivery: IP to your machine via SLIP

Documentation:

Minimum hardware/
software requirements: Mac or PC with Windows, plus modem (9600 baud recommended)

Fees: PC Windows or Mac versions £180/year + £50 initial set-up fee (This includes one e-mail Post Office Protocol connection; more can be accessed from the same IP Gold account for an additional £50/year each).
No on-line or other charges
(+VAT)

For more information:

Tel:	0223 566950
Fax:	0223 566951
E-mail:	sales@cityscape.co.uk

Comments: Based around the basic 'Mail-it' package for e-mail, Mosaic for Web and other forms of interactive access.
See also CityScape E-mail service.

CIX (Compulink Information eXchange)*

Address:

Compulink Information eXchange Ltd,   *or*   The Sanctuary,
London House,   Oakhill Grove,
Ancaster Square,   Surbiton,
Llanrwst,   Surrey KT6 6DU
Gwynedd LL26 0LD

POPs: London

Proposed POPs
(by end 1994):

Basic services: E-mail, telnet, FTP (from host machine)

Additional services: USENET News, FTP space, conferencing, outgoing Fax

Help/Support:

Method of connection: Dial-up, ISDN, X.25, Internet

Form of service delivery: Terminal access, off-line reader

Documentation: User manual (one with registration fee, additional copies £10)

Minimum hardware/
software requirements: Any computer with a basic communications package and modem or access to X.25 services, ISDN or the Internet

Fees: £25 registration fee Monthly minimum charge of £6.25 Connection rates – £3.20/hour peak, £2.40/hr off-peak Faxes extra (+VAT)

For more information:

Tel:	081 390 8446 or 0492 641961
Fax:	081 390 6561
E-mail:	cixadmin@compulink.co.uk

Comments: There is a Windows based off-line reader called
Ameol – £45 + VAT.

*CIX (Compulink Information eXchange)
should not be confused with CIX (Commercial
Internet eXchange). The latter is a cooperative
grouping of the major commercial IP network
providers in the US and the rest of the world
who have agreed to interwork their networks.

CompuServe

Address: CompuServe Information Service,
1 Redcliff Street, PO Box 676, Bristol BS99 1YN

POPs: Bristol, Reading, London, Manchester,
Birmingham, Edinburgh (*see* Comments below)

Proposed POPs
(by end 1994):

Basic services: E-mail (from host machine)

Additional services: Range of bulletin boards (known as forums)
including some for hardware/software support,
and so on. Some information services are free
(part of the basic package) but many involve
an extra fee or fee plus supplement.

Help/Support:

Method of connection: Dial-up (GNS Dialplus and Mercury 5000
access available at extra cost)

Form of service delivery: Terminal access

Minimum hardware/
software requirements: Any computer that can run a communications
package and connect to a modem.
CompuServe recommend their *Information
Manager* graphical interface package for PCs
(CIM or WinCIM) and Macs (included in
membership kit).

Fees: £24.95 to join (includes software package) +
$8.95/month + daytime (Mon-Fri 8:00 – 19:00)
connect charge $7.70/hr + charges for
incoming Internet mail + extra for non-
standard services. Use of CompuServe's own

network is free evenings and weekends, other times there is a charge. There is always a charge for the Mercury or GNS access services which varies according to the time of day, etc. All prices except joining fee in US Dollars so what you pay varies with the exchange rate. Prices include VAT.

For more information:

Tel: 0800 289458
Fax: 0272 252210
E-mail: 70006.101@compuserve.com

Comments:

CompuServe pre-dates the public availability of the Internet (it was established in 1979). Currently its only link to the Internet is via e-mail. It intends to extend its Internet services in late 1994, first to make USENET Newsgroups available, later to include FTP and other interactive services.

Delphi Internet Ltd

Address:

Delphi Internet Limited, The Elephant House, Hawley Crescent, London NW1 8NP

POPs:

London

Proposed POPs
(by end 1994):

Basic services:

E-mail, FTP, telnet, local forums (conferences) on host machine

Additional services:

USENET News, Gopher, WWW, software archives, access to a database of *The Times* and *Sunday Times*, IRC, games, and so on.

Help/Support:

Customer Service desk open 10 a.m. – 10 p.m. weekdays and 12 – 7 p.m. weekends

Method of connection:

Dial-up direct (up to 14400 baud), via GNS DialPlus (up to 2400 baud) or telnet

Form of service delivery:

Terminal access. Offline readers and graphical navigators available soon

Documentation:

Online help, New Member Welcome Booklet, 290-page book ('Delphi: The Official Guide')

Minimum hardware/ software requirements:	Any machine that can run terminal emulation software plus a suitable modem
Fees:	10/4 Plan – £10/month. This gives 4 hours 'free' on-line time per month then £4/hr. 20/20 Plan – £20/month. This gives 20 hours 'free' on-line time per month then £1.80/hr Surcharge for access via BT's GNS service – £1.50/hr (+VAT)
For more information:	*Tel:* 071 757 7080 *Fax:* 071 757 7160 *E-mail:* ukservice@delphi.com or uk@dephi.com
Comments:	Delphi is a large US on-line service which has been recently launched in the UK.

Demon Internet

Address:	Demon Systems Ltd, 42 Hendon Lane, London N3 1TT
POPs:	London, Warrington (local call from Liverpool and Manchester), Edinburgh, Sunderland (local call from Durham and Newcastle), Reading, Leeds, Sheffield, Hull and Bradford
Proposed POPs (by end 1994):	Birmingham, Cambridge and others
Basic services:	E-mail feed, FTP, telnet (from your machine)
Additional services:	USENET news feed, on-line readers, IRC, public FTP archive, 'dial on demand', sub- domain addresses, and so on.
Help/Support:	
Method of connection:	Dial-up, dedicated/reserved line, leased line
Form of service delivery:	TCP/IP (SLIP/PPP for dial-up)
Documentation:	
Minimum hardware/ software requirements:	PC, Mac, Amiga, UNIX machines (and others).

Appropriate free software can be provided

Fees: Dial-up £12.50 registration, £10/month, no
time charges.
Dedicated/Reserved 14,400 line £750 set-up +
£100/month
Leased 14.4K line £1000 set-up + £200/month
Leased 64K line £1000 set-up + £400/month
Other services available
(+VAT)

For more information: *Tel:*   081 349 0063   081 343 3881
Fax: 081 349 0309
E-mail: internet@demon.net
FTP: ftp.demon.co.uk:/pub/doc/Services.txt

Comments: The above is only a selection – Demon provide
a wide range of services from the connection of
a single machine to support for a complete
sub-network. They also supply a limited range
of Internet and Unix related books.

Direct Connection

Address: The Direct Connection Ltd, PO Box 931,
London SE18 3PW

POPs: London

Proposed POPs
(by end 1994):

Basic services: telnet, FTP, E-mail (from your or host machine)

Additional services: USENET (multiple on-line readers), Gopher,
Archie, FAX gateway, Hytelnet, FTP space,
Batch FTP, UUCP (mail/news) feeds, Archie
and Gopher clients

Help/Support:

Method of connection: Dial-up

Form of service delivery: Terminal access, or TCP/IP (SLIP/PPP), or UUCP

Documentation:

Minimum hardware/ software requirements:	For terminal access – any computer than can run a standard communications and terminal emulation package. For TCP/IP – most machines that can run TCP/IP (PC, Mac, UNIX boxes, etc). For UUCP – most larger micros and all UNIX machines. All services need a modem.
Fees:	From £10/month (+ VAT), unlimited use, no time charges
For more information:	*Tel:* 081 317 0100 *E-mail:* helpdesk@dircon.co.uk *Dial-up:* 081 317 2222 – login as 'demo' for information or sign-up
Comments:	A good range of 'products' – both terminal access (easy to get started) and TCP/IP direct from your own computer, plus UUCP feeds for those who want it.

Direct Line (Powerline Systems Ltd)

Address:	NOT KNOWN
POPs:	London
Proposed POPs (by end 1994):	
Basic services:	E-mail (from host machine)
Additional services:	USENET, support for off-line readers, large software archive (DOS, Windows, OS2, Mac, and others)
Help/Support:	
Method of connection:	Dial-up
Form of service delivery:	Terminal access (+ off-line readers)
Documentation:	
Minimum hardware/ software requirements:	Any computer that can run a basic communications and terminal emulator package + modem

Fees: £25 p.a.(+ VAT)

For more information: *Tel:* Not known
Fax: 081 845 8952
E-mail: sysop@ps.com

Comments: Basic information from network lists. It has proved difficult to get further information on this company.

Edex (The Education Exchange)

Address: The Education Exchange, Galviz Ltd, 3 Stroud Road, Wimbledon Park, London SW19 8DQ

POPs: London, Leeds and Aberdeen
(NB: *see Fees* below)

Proposed POPs
(by end 1994):

Basic services: E-mail, FTP, telnet (host or own machine)

Additional services: USENET, Local databases (text and software), Gopher, Web, WAIS (client software on host or for own machine)

Help/Support:

Method of connection: Dial-in (to Edex or GNS service), X.25

Form of service delivery: Terminal access, UUCP, or PPP I/P access

Documentation: User guide

Minimum hardware/
software requirements: Any computer with a basic communications package and modem

Fees: Basic service (E-mail/USENET) – £50/year
Interactive service – £70/year

+ Connection charges:
London direct dial – None

Outside London:
– Direct dial – None
– Dial-in via Leeds or Aberdeen POP – £0.60/hr
– Dial in via Edex's GNS account – £2.20/hr
(+VAT)

For more information: *Tel:*   081 944 8021
                        Fax: 081 944 5029
                        Dial-up: 081 944 8026

Comments: Edex's approach is to offer basic terminal
            access services to first-time users, but they
            also have a wide range of other host based and
            user machine based services available on
            request. Edex is intending to construct its own
            2 Mbit network which will be available in early
            1995. Edex is aimed at schools but does offer
            individual subscriptions.

ElectricMail

Address: ElectricMail Ltd Orwell House, Cowley Road,
           Cambridge CB4 4WY

POPs: As EUnet (*see* Comments below)

Proposed POPs
(by end 1994):

Basic services: E-mail, FTP and telnet (reselling EUnet
                  services from your machine)

Additional services: Consultancy, corporate network design for e-
                       mail, fax and conferencing systems,
                       interworking Internet and LANs, network
                       security, and so on. Wide range of e-mail
                       gateway software to enable Internet mail to
                       cc:mail, MS mail, Novell MHS mail, and others.
                       Various packaged solutions. Provision of
                       hardware and software as required.

Help/Support: Operational system support, remote network
                monitoring, and so on.

Method of connection: Dial-in, ISDN, leased line, and so on.

Form of service delivery: UUCP, TCP/IP (leased line or SLIP, PPP), etc

Documentation:

Minimum hardware/
software requirements: As for EUnet

Fees:	Fees are as EUnet when reselling their services. Some packaged solutions, otherwise usual consultancy arrangements.
For more information:	*Tel:* 0223 420193 *Fax:* 0223 420195 *E-mail:* info@elmail.co.uk
Comments:	This company is a little different from most of the others here as it does not, itself, provide network access but instead resells EUnet services. Its main work is consultancy and systems design/integration. The only other company of this type listed here is Motiv Systems Ltd.

EUnet GB Ltd

Address:	EUnet GB, Kent R&D Business Centre, Giles Lane, Canterbury, Kent, CT2 7PB, UK
POPs:	London, Bracknell, Cambridge, Canterbury, Birmingham, Glasgow
Proposed POPs (by end 1994):	Bristol, Belfast, Aberdeen, Swindon, Southampton
Basic services:	E-mail, FTP, telnet, TCP/IP (from your machine)
Additional services:	UUCP, E-mail feed, news feed, Fax gateway, Lotus cc:mail gateway, MHS: mail gateway, Pan-European VPN (Virtual Private Network), on-site configuration/network set-up
Help/Support:	9 a.m. - 9 p.m. Mon-Fri – telephone, fax and e-mail
Method of connection:	Dial-up, ISDN, Leased-line, X.25
Form of service delivery:	TCP/IP over leased line, UUCP or SLIP/PPP over dial-up, IP over X.25
Documentation:	
Minimum hardware/ software requirements:	Any machine that can run UUCP or SLIP/PPP + modem. Alternatively, a hardware router based connection

Fees:	UUCP 'EmailLink' – £95/qtr
	UUCP 'EmailLink + News' – £145/qtr
	IP Dial from £450/qtr + £300 set-up
	ISDN from £750/qtr + 450 set-up
	Leased line from £1250/qtr + £1000 set-up
	Additional fees if equipment is leased from EUnet (+VAT)
For more information:	*Tel:* 0227 475497
	Fax: 0227 475478
	E-mail: sales@Britain.EU.net
Comments:	EUnet GB (previously UKnet) is part of EUnet which provides networking services across Europe (including central and eastern Europe) and North Africa.

ExNet Systems Ltd

Address:	ExNet Systems Ltd, 37 Honley Road, Catford, London SE6 2HY
POPs:	London, Edinburgh
Proposed POPs (by end 1994):	
Basic services:	E-mail, telnet and FTP
Additional services:	E-mail feed (UUCP), USENET News feed, FTPmail, batched FTP, Gopher, and so on.
Help/Support:	
Method of connection:	Dial-up (ISDN and leased line possible)
Form of service delivery:	Terminal access, UUCP, IP (PPP)
Documentation:	Manual (£13)
Minimum hardware/ software requirements:	
Fees:	Basic services (E-mail and News) from £84/year
	Basic + UUCP services from £108/year
	Internet interactive services from host £120/year*

Internet interactive service from users machine
(PPP) from £240/year* – other options
available. *as supplement to basic services.
Above prices assume modem connection,
other forms of connection (ISDN, leased line)
at additional cost.
(+VAT)

For more information:
   Tel: 081 244 0077
   Fax: 081 244 0078
   E-mail: Helpex@exnet.com

Comments:
Originally a dial-up terminal access and/or
UUCP e-mail + USENET News service. Now also
offers a range of Internet interactive services
from their host machine and/or your machine.

Genesis Project Ltd

Address:
Genesis Project Ltd, ITCB Interpoint,
20-24 York Street, Belfast, BT15 1AQ

POPs:
Belfast

Proposed POPs
(by end 1994):
Dublin and Cork

Basic services:
E-mail, FTP, telnet (from your machine)

Additional services:
FTP and WWW server space available to third
parties. Consultancy for LAN and WAN
management.

Help/Support:
24-hour for leased line services

Method of connection:
Dial-up, ISDN (dial-up and reserved connections),
reserved (dedicated) line, leased line

Form of service delivery:
IP from your machine over dial-up, reserved
connection, or leased line.

Documentation:

Minimum hardware/
software requirements::
Any machine that can run serial line IP
software + modem. Alternatively, a hardware
router based connection.

Fees:
Dial-up (single machine) – £10/month +
£12.50 to join

ISDN (single machine) – £180/month +
£150 to join
Dial-up/Reserved line – 14.4K – £100/month +
£750 set-up
Leased line – 64K – £400/month + £1000
set-up
Other options available
(+VAT)

For more information:	*Tel:*	0232 231715
	Fax:	0232 231622
	E-mail:	info@gpl.com
	FTP:	ftp.gpl.net

Comments: Genesis is a general IT company doing work in the areas of e-publishing and multimedia. The Internet access service is provided by its Communications Division. It provides this service in partnership with Unipalm/PIPEX.

GreenNet

Address:	GreenNet, 4th Floor, 393-395 City Road, London EC1V WE
POPs:	London
Proposed POPs (by end 1994):	
Basic services:	E-mail, telnet (from host machine)
Additional services:	Conferences, WAIS, USENET News, indirect FTP, Fax
Help/Support:	
Method of connection:	Dial-up, X.25, BT Dialplus, or telnet (Internet)
Form of service delivery:	Terminal access
Documentation:	
Minimum hardware/ software requirements:	Any computer that can run a basic communications and terminal emulator package + modem or X.25 access
Fees:	Non-commercial use: £15 registration, £5/month + connect charges of £3.60/hr peak or £2.40/hr off-peak

Commercial use:
£30 registration, £10/month + connect
charge of £6/hr
Access via GreenNet's DialPlus account –
£1.80/hr extra (+VAT).

For more information: *Tel:* 071 608 3040
         Fax: 071 253-0801
         E-mail: support@gn.apc.org

Comments: As implied by its name, GreenNet is involved
in the support of many environmental and
human rights organisations. It is the UK
member of APC (Association for Progressive
Communications). For further information on
APC see Appendix E.

HEAnet (Higher Education Authority network)

Address: HEAnet, 21 Fitzwilliam Square, Dublin 2,
Ireland

POPs: Dublin

Proposed POPs
(by end 1994): Cork, Limerick, Galway

Basic services: E-mail, FTP, telnet (from your machine)

Additional services: Support for DECnet and Novell connections,
National Archive Server, X.400 gateway, X.500
directory, X.25 within HEAnet and EuropaNET

Help/Support: Help Desk

Method of connection: Leased line

Form of service delivery: TCP/IP to your machine

Documentation:

Fees: For new users and 64K service:
IE£2,000 set-up charge, IE£4,000 for first year.
Then IE£8,344 per year.
Many other levels of service available.

For more information: *Tel:* +353 (0)1 6612748
         Fax: +353 (0)1 6610492
         E-mail: mnorris@hea.ie

Comments: HEAnet is an academic network similar in many ways to JANET in the UK. It has an Acceptable Use Policy that excludes straight commercial use but may allow the connection of organisations involved in collaboration with academic or research activities, or organisations that support academic activities. For details on exactly who can and cannot connect contact Mike Norris (details above).

NB: HEAnet does not support an equivalent to JANET's 'secondary' services (see the entry for JANET for an explanation of 'secondary' services).

IEunet

Address: IEunet Ltd, Innovation Centre, Trinity College, Dublin 2, Ireland

POPs: Dublin, Galway, Shannon

Proposed POPs (by end 1994): Cork

Basic services: E-mail/USENET News (UUCP) or Full TCP/IP (E-mail, FTP, telnet, and so on) – from your machine

Additional services: USENET News, FTP/Mail server, X.400 mail, Fax out, incoming e-mail to fax, X.25 service, communications hardware supply, and so on.

Help/Support:

Method of connection: Dial-up, leased line

Form of service delivery: UUCP for EMailLink, and so on. SLIP for IPLink, or full TCP/IP over leased line

Documentation:

Minimum hardware/ software requirements: Any machine that can run UUCP for E-mail/News, or SLIP for dial-up IP, or full TCP/IP for leased line + modem or router

Fees: EMailLink (UUCP) – Fees range from IE£25/month for up to 1Mbyte /month to IE£250/month for over 13Mbyte/month +

IE£100 registration EMailLink+News – News is
IE£35/month extra to EMailLink charges
(full feed, approximately 80Mbytes/day)
DIAL-IP – Registration IE£100 + IE£8/hour
on-line charge (min IR£24/month), includes
E-mail and News over IP
IPLine – IE£5000/year + IE£1000 set up (does
not include line rental or router). Router –
supplied, configured and installed – IE£2000.
Other options available (+VAT)

For more information:	*Tel*:	+353 (0)1 679 0832 or 671 9361
	Fax:	+353 (0)1 679
	E-mail:	info@ieunet.ie

Comments: Part of EUnet. *See* also EUnet (GB).

INFONET Services

Address:	INFONET Services, P.O. Box 148, Corcaigh, REPUBLIC OF IRELAND
POPs:	Cork
Proposed POPs (by end 1994):	
Basic services:	E-mail (from host machine) and Bulletin Board based services
Additional services:	FTPMAIL, Virtual 'private' BBs and e-mail information server, off-line mail reader, menu front-end, software archive, communications and LAN consultancy, and so on.
Help/Support:	
Method of connection:	Dial-up
Form of service delivery:	Terminal access (with or without off-line reader)
Documentation:	
Minimum hardware/ software requirements:	Any computer that can run a basic communications and terminal emulator package + modem

Fees:	The following assume a maximum of 2hr/day + 2 Mbyte download from BB.
	Casual e-mail user – less than 256K/month mail – IE£75/year
	Basic Professional e-mail user – less than 1M/month mail IE£145/year
	Ordinary Professional – as Basic Professional + company address alias IE£195/year
	Other services/options available (+VAT)
For more information:	*Tel*: +353 (0)21 293593 or (0)88 596607
	Fax: +353 (0)21 295574
	E-mail: office@infonet.ie
	Dial-up: +353 (0)21 294914
Comments:	A Bulletin Board service and FidoNet system offering e-mail access to the Internet.

Ireland On-Line

Address:	Ireland On-Line, West Wing, Furbo, Galway, Ireland
POPs:	Dublin, Galway
Proposed POPs (by end 1994):	Limerick, Cork
Basic services:	E-mail, FTP, telnet (from host machine), e-mail/USENET/News (from your machine). Other services possible – *see* Comments below.
Additional services:	ClariNet, local discussion groups, software archive, FTP/Gopher/WWW servers for third party use, off-line reader (IOL-Win client), provision of communications hardware and software, consultancy, and so on.
Help/Support:	Full-time support. Training and installation available
Method of connection:	Dial-up (but *see* Comments)
Form of service delivery:	Terminal access or UUCP (for E-mail/USENET News)

Documentation:

Minimum hardware/
software requirements: Any computer that can run a basic
communications and terminal emulator
package or UUCP (+ modem)

Fees: Basic service (E-mail/USENET News, local
conferencing, Web and other clients) –
IE£10/month (20 hours/month) or IE£20/month
(unlimited use) + IE£25 registration.
Premium services (telnet, FTP, and so on) –
IR£3/hour extra.
Group discounts available – other options
available.
(+VAT)

For more information: *Tel*: +353 (0)91 92727
Fax: +353 (0)91 92726
E-mail: info@iol.ie
FTP: ftp.iol.ie
Dial-up: +353 (0)1 285 2700 (Dublin)
+353 (0)91 92711 (Galway)

Comments: Ireland On-Line can also supply leased line
and dial-up IP services in conjunction with
IEunet.

JANET (Joint Academic Network)

Address: UKERNA[1], Atlas Centre, Chilton, Didcot,
OXON OX11 0QS

POPs: UK wide (*see Comments* below)

Proposed POPs
(by end 1994):

Basic services: All possible – depends on form of connection –
see Comments below

Additional services:

Help/Support:

Method of connection: Primary – leased line (IP, X.25, IP over X.25, and
so on). Secondary /Affiliated – Full (as Primary),
Restricted – Dial up, ISDN, X.25, and so on.

Form of service delivery:	Depends on form of connection – see Comments below
Documentation:	
Minimum hardware/ software requirements:	Depends on form of connection – see Comments below
Fees:	Depends on form of connection – see Comments below
For more information:	*Tel*: 0235 445517
	Fax: 0235 446251
	E-mail: janet-liaison-desk@jnt.ac.uk
Comments:	JANET is an academic/research network and operates according to *Acceptable Use Guidelines*. To be eligible to connect you must be an academic or research organisation, or collaborating in some way with such an organisation, or an organisation whose connection will be 'beneficial to a significant sector of the academic community'. Full (primary) connections are direct to the network but there are now 'secondary' and 'affiliated'[2] connections which are made via an organisation with a primary connection (for example, a University).

[1] United Kingdom Education and Research Network Association (Until 1994 known as the JNT (Joint Network Team).

[2] 'Affiliated' connections are technically identical to 'secondary' connections but only available to publicly funded colleges of further or higher education.

Kirklees Host

Address:	Kirklees Host, Field House, Wellington Road, Dewsbury WF13 1HF
POPs:	Kirklees
Proposed POPs (by end 1994):	
Basic services:	E-mail, telnet from host machine and POP4 e-mail to your machine
Additional services:	Telex, Fax, X.400 e-mail, 'Alert' service, bulletin boards and conferences, on-host and external database access, private/public database provision Web Server, off-line readers
Help/Support:	Help desk available during office hours. Training programmes in London and Manchester
Method of connection:	Dial-up, X.25, GNS Dial Plus (including use of access providers account)
Form of service delivery:	Terminal access, and/or POP4 for e-mail
Documentation:	
Minimum hardware/ software requirements:	Any computer that can run a basic communications and terminal emulator package + modem or X.25 access
Fees:	Registration fee – £25 E-mail/telnet – £10/month + additional charge during peak times of 5p/minute Fax, Telex and remote database access charged by use. (+VAT)
For more information:	*Tel*: 0924 457070 *Fax*: 0924 457072 *E-mail*: kirklees-host@geo2.geonet.de *X.400*: c=de a=dbp p=geonet o=softsolution u=kirklees
Comments:	All the Host services are aimed at providing IT solutions for small to medium business users and the not-for-profit sector (voluntary organisations, trade unions, and so on). One of the 'Host' systems set up by Soft Solution Ltd (see POPTEL below). Part of GeoNet.

London Host (previously known as the Hackney Host)

Address:	London Host, 90 De Beauvoir Road, London N1 4EN
POPs:	London
Proposed POPs (by end 1994):	
Basic services:	E-mail
Additional services:	Telex, e-mail to fax, fax to e-mail (as image) bulletin boards and conferences, X.400 e-mail, on-host and external database access, VT100 Web client, telnet
Help/Support:	
Method of connection:	Dial-up, X.25
Form of service delivery:	Terminal access
Documentation:	
Minimum hardware/ software requirements:	Any computer that can run a basic communications and terminal emulator package + modem or X.25 access
Fees:	Registration fee – £25 E-mail/telnet – £10/month (includes £50/month 'free' on-line time) + additional charges of 12p/min (peak) Fax, Telex and remote database access charged by use (+VAT)
For more information:	*Tel*: 071 241 1000 *Fax*: 071 242 5007 *E-mail*: sales@lond.geonet.de *X.400*: c=de a=dbp p=geonet o=computer-access
Comments:	In partnership with the Manchester and Kirklees Hosts but in this case set up by Computer Access. All the Host services are aimed at providing IT solutions for small to medium business users and the not-for-profit sector (voluntary organisations, trade unions, and so on). Part of GeoNet.

Manchester Host

Address:	Manchester Host, 30 Naples Street, Manchester M4 4DB
POPs:	Manchester,
Proposed POPs (by end 1994):	
Basic services:	E-mail, telnet from host machine and POP4 e-mail to your machine
Additional services:	Telex, Fax, bulletin boards and conferences, on-host and external database access, private/public database provision Web Server, off-line readers
Help/Support:	Help desk available during office hours. Training programmes in London and Manchester
Method of connection:	Dial-up, X.25, GNS Dial Plus (including use of access providers account)
Form of service delivery:	Terminal access, and/or POP4 for e-mail
Documentation:	
Minimum hardware/ software requirements:	Any computer that can run a basic communications and terminal emulator package + modem or X.25 access
Fees:	Registration fee – £25 E-mail/telnet – £10/month + additional charge during peak times of 5p/minute Fax, Telex and remote database access charged by use (+VAT)
For more information:	*Tel*: 061 839 4212 *Fax*: 061 839 4214 *E-mail*: admin@mcr1.poptel. org.uk
Comments:	All the Host services are aimed at providing IT solutions for small to medium business users and the not-for-profit sector (voluntary organisations, trade unions, and so on). One of the 'Host' systems set up by Soft Solution Ltd (see poptel below). Part of GeoNet.

Motiv Systems Ltd

Address:	Motiv Systems Ltd, 22 Hills Road, Cambridge CB2 1JP
POPs:	As Demon Internet (see Comments below)
Proposed POPs (by end 1994):	As Demon Internet
Basic services:	E-mail feed, FTP, telnet (reselling Demon Internet services)
Additional services:	As Demon Internet + consultancy services for access to the Internet, Internet tool selection, development of custom Internet applications for DOS, Unix, Novell, Mac and other platforms, assistance with Internet publication + other network consultancy services
Help/Support:	
Method of connection:	dial-up, dedicated modem, leased line
Form of service delivery:	TCP/IP (SLIP/PPP for dial-up)
Documentation:	
Minimum hardware/ software requirements:	PC, Mac, Amiga, UNIX machines (and others). Appropriate free software can be provided.
Fees:	As Demon Internet for Internet access, other fees as consultancy
For more information:	*Tel*: 0223 576318 *Fax*: 0223 576319 *E-mail*: PaulR@Motiv.demon.co.uk
Comments:	This company is a little different from most of the others here as it does not, itself, provide network access but instead resells Demon Internet services. Its primary work is Open Systems and Internet consultancy. The only other company like this in this list is ElectricMail.

On-line Entertainment Ltd

Address:	On-line Entertainment Ltd, 642a Lea Bridge Road, London, E10 6AP
POPs:	London
Proposed POPs (by end 1994):	
Basic services:	E-mail, FTP, telnet (on the host machine)
Additional services:	On-line multi-player games, BBS, USENET News, on-line readers, chat system
Help/Support:	
Method of connection:	Dial-up or dial-in via X.25 access. Also Internet telnet access.
Form of service delivery:	Terminal access
Documentation:	
Minimum hardware/ software requirements:	Any computer that can run a basic communications and terminal emulator package + modem or X.25 access
Fees:	All services including access to Internet – £2.00/hr Access via On-Line's DialPlus/Tymnet account – £1.59/hr extra (+VAT)
For more information:	*Tel*: 081 558 6114
	Fax: 081 558 3914
	E-mail: jon@on-line.co.uk
Comments:	Currently a mainly entertainment service but moving towards offering more business oriented services in the future.

Pavilion Internet plc

Address:	24 Old Steine, Brighton, E. Sussex, BN1 1EL
POPs:	Brighton, London, Bristol, Birmingham, Manchester, Edinburgh
Proposed POPs (by end 1994):	Portsmouth, London (Cable), Manchester (Cable)
Basic services:	E-mail, FTP, Telnet, WWW, IRC, News (from your machine). Configured software provided for Macintosh or PC/Windows 3.
Additional services:	Domain name registration, Space on WWW/FTP server
Help/Support:	10–6 Mon–Fri by Phone/Fax, 6–10 on site within 30 miles of Brighton (extra charge), e-mail 'any time'
Method of connection:	Dial-up (50% of lines leased from NYNEX Cablecomms allowing free local calls to NYNEX customers)
Form of service delivery:	IP via SLIP/PPP over dial-up
Documentation:	Starter guide
Minimum hardware/ software requirements:	Recommend PC with Windows 3 or Mac. Otherwise any machine that runs SLIP. Most support functions use WWW, so Mosaic or another Web client is advisable.
Fees:	£17.75 Registration £17.75/month (£14.75 if paid by Direct Debit) No connection charges (Includes VAT)
For more information:	*Tel*: (01273) 607072 *Fax*: (01273) 607073 *E-mail*: info@pavilion.co.uk *Web*: http://www.pavilion.co.uk/
Comments:	Aiming to make the Internet 'accessible to the person in the street'. Software provided pre-configured. Dedicated to alternative uses of the Internet, and to that end are working on a number of arts-based services, galleries and so on. Interesting use of cable connection.

PC User Group CONNECT service

Address:	PC User Group Ltd, PO Box 360, 84-88 Pinner Road, HARROW HA1 4LQ
POPs:	London
Proposed POPs (by end 1994):	
Basic services:	E-mail, FTP, telnet (from host machine)
Additional services:	BBS, IRC, USENET News feeds, public FTP Archive, movie database server (movie@ibmpcug.co.uk), Gopher, UUCP, Batch FTP, games, and so on.
Help/Support:	
Method of connection:	Dial-up, X.25, and telnet access
Form of service delivery:	Terminal access, UUCP
Documentation:	
Minimum hardware/ software requirements:	Any computer that can run a basic communications and terminal emulator package + modem or X.25 access
Fees:	(Non members of PC Usergroup) Registration fee – £10 Host based (depending on range of services chosen) – from £5 to £14/month or from £50 to £175/yr. UUCP (site/multi-user access) – Mail only – £250/yr – News and Mail – £400/yr No time charges (+VAT)
For more information:	*Tel*: 081 863 1191 *Fax*: 081 863 6095 *E-mail*: info@ibmpcug.co.uk *FTP*: ftp.ibmpcug.co.uk:/pub/doc/services *Dial-up*: 081 863 6646 (Login as 'register' and enter '?' at the 'nickname' prompt)
Comments:	The PC User Group has been providing similar services for its members since 1989. It has over 10,000 members. Fees are reduced for members. See also the separate entry for WinNET Mail & News service – also provided by the PC User Group.

PIPEX

Address:	Pipex Ltd, 216 Cambridge Science Park, Milton Road, Cambridge, CB4 4WA
POPs:	Leased line and dial-up: London, Cambridge, Edinburgh. Dial-up (being upgraded to leased line soon): Birmingham, Bristol, Manchester
Proposed POPs (by end 1994):	
Basic services:	E-mail, telnet, FTP (from your machine)
Additional services:	USENET news feed, Windows e-mail reader, X.400 ADMD, batch FTP, indirect FTP, FTP space, PAD, telnet login, fax gateway, Gopher, WAIS, WWW client, WWW server, anon FTP, Class B & Class C registrations, Domain name registration, consultancy, routers, modems, encryption, network management, and so on.
Help/Support:	
Method of connection:	Dial-up, ISDN, X.25, leased line
Form of service delivery:	Full TCP/IP (SLIP, PPP for dial-up), NNTP Newsfeed
Documentation:	
Minimum hardware/ software requirements:	Any computer with appropriate software and communications hardware
Fees:	Dial-up (PSTN) 'Caller' service – £2000/yr + £250 set-up Dial-up (X.25) 'Caller' service – £2000/yr + £250 set-up Dial-up (ISDN) – (Sun machines) £4000/yr + £1000 set-up Leased line (64K) 'UK' service – £7500/yr + £1000 set-up Leased line (64K) 'World' service – £9600/yr + £1000 set-up None of the above include communications hardware, which can be leased at extra cost. Many other options (+VAT)

For more information: *Tel*:   0223 250120
                              Fax: 0223 250121
                              E-mail: sales@pipex.net
                              X.400:   C=GB;A=PIPEX;S=PIPEX
                              FTP: ftp.pipex.net/pub/FAQ

Comments: PIPEX offer a wide range of services aimed mainly at the corporate market.

Poptel

Address: POPTEL/Soft Solution, 25 Downham Road, London N1 5AA

POPs: London

Proposed POPs (by end 1994):

Basic services: E-mail, telnet from host machine and POP4 e-mail to your machine

Additional services: Telex, Fax, X.400 e-mail, 'Alert' service, bulletin boards and conferences, on-host and external database access, private/public database provision, private/public database provision Web Server, off-line readers

Help/Support: Help desk available during office hours. Training programmes in London and Manchester

Method of connection: Dial-up, X.25, GNS Dial Plus (including use of access providers account)

Form of service delivery: Terminal access, and/or POP4 for e-mail

Documentation:

Minimum hardware/ software requirements: Any computer that can run a basic communications and terminal emulator package + modem or X.25 access

Fees: Registration fee – £25
E-mail/telnet – £10/month + additional charge during peak times of 5p/minute
Fax, Telex and remote database access charged by use
(+VAT)

For more information:	*Tel*:	071 249 2948
	Fax:	071 254 1102
	E-mail:	poptel-admin@mcr1.poptel.org.uk

Comments: Poptel was set up by Soft Solution Ltd who later set up the Manchester and Kirklees hosts. All the Host services are aimed at providing IT solutions for small to medium business users and the not-for-profit sector (voluntary organisations, trade unions, and so on). Poptel is part of GeoNet.

RedNet onLine

Address: Rednet (onLine) Ltd, 6 Clivedon Office Village, High Wycombe, BUCKS HP12 3YZ

POPs: Birmingham, Bristol, Cambridge, Edinburgh, London and Manchester

Proposed POPs
(by end 1994):

Basic services: Dial-up IP from your machine

Additional services: PC users: E-mail (using POP3), USENET News, e-mail, and Web access. Mac users: e-mail, FTP, USENET News and Web access. Space on FTP and Web servers. IRC server. Mail forwarding option allows tailoring of a user's e-mail address. Other services including hardware supply, consultancy and training.

Help/Support:

Method of connection: Dial-up

Form of service delivery: IP via SLIP or PPP

Documentation:

Minimum hardware/
software requirements: PC or Mac plus modem

Fees: Basic account (includes POP3 mail service) – £15/month, start-up £25 POP3 mail account (for existing Internet users with IP access) – £7.50, start-up £15
Mail forwarding option – £120/year
(+VAT)

For more information: *Tel*:   0494 513333
                          Fax: 0494 443374
                          E-mail: support@rednet.co.uk
                          Web: http://w3.rednet.co.uk

Comments: PC users are supplied with the basic TCP/IP
              software as part of the package but Mac users
              need to buy this separately.

Sound & Vision BBS

Address: Sound & Vision, 24 Oatlands Chase,
             Weybridge, SURREY KT13 9RY

POPs: Weybridge (local call access from London,
          Guildford, Slough and Staines)

Proposed POPs
(by end 1994):

Basic services: E-mail (from host machine)

Additional services: BBS, USENET, Games, software on CD-ROMs,
                         chat

Help/Support:

Method of connection: Dial-up

Form of service delivery: Terminal access to BB, Mail/News download by
                              proprietary software

Documentation:

Minimum hardware/
software requirements: Any computer that can run a basic
                           communications and terminal emulator package
                           + modem

Fees: E-mail/News + minimal BB access – £15/year
          E-mail/News + 2 hour/day BB access – £30/year
          E-mail/News + unlimited BB access – £60/year
          No time/kbyte charges
          (+VAT)

For more information: *Tel*:   0932 253131
                          E-mail: rob@span.com
                          Dial-in: 0932 252323

Comments: A large Bulletin Board service with over 3000
              members. Currently operating with 6 lines but
              upgrading soon to 25 lines.

Specialix International

Address:	Specialix International, 3 Wintersells Rd, Byfleet, Surrey KT14 7LF
POPs:	Byfleet
Proposed POPs (by end 1994):	
Basic services:	USENET News feed
Additional services:	None
Help/Support:	
Method of connection:	Dial-up
Form of service delivery:	UUCP or NNTP
Documentation:	
Minimum hardware/ software requirements:	Any computer that can run a basic communications package and UUCP or NNTP + modem
Fees:	Full USENET News feed – £200/year
For more information:	*Tel*: 0932 354254 *Fax*: 0932 352781 *E-mail*: postmaster@specialix.co.uk *FTP*: ftp.specialix.co.uk:/public/news.service
Comments:	This is a USENET News feed only – no e-mail or other services. All known groups are carried.

Sprint International – Global SprintLink Service

Address:	Sprint International, Rawdon House, Bond Close, Kingsland Business Park, Basingstoke RG24 0PZ
POPs:	(UK) London (others in major cities around the world)
Proposed POPs (by end 1994):	
Basic services:	Leased line IP service
Additional services:	Gateways to Sprint's Global Frame Relay Service (GFRS), SprintNet (X.25) and SprintMail (X.400 e-mail)
Help/Support:	24-hr/7-day Hotline
Method of connection:	Leased line
Form of service delivery:	IP to user's machine or network
Documentation:	
Minimum hardware/ software requirements:	Any computer that can support TCP/IP connectivity plus appropriate communications hardware
Fees:	Not known
For more information:	_Tel_: 0256 811181 (Robin Loxley) _Fax_: 0256 811149 _E-mail_: Robin.Loxley@sprintintl.sprint.com
Comments:	An international service with a UK node. Aimed mainly at the corporate sector plus Government and Internet access resellers.

Spud's Xanadu

Address:	NOT KNOWN
POPs:	Coventry
Proposed POPs (by end 1994):	
Basic services:	E-mail (from host machine)
Additional services:	Menu front-end, USENET News, on-line readers, FTP mail, FAX gateway, BBS, games, chat system
Help/Support:	
Method of connection:	Dial-up
Form of service delivery:	Terminal access
Documentation:	
Minimum hardware/ software requirements:	Any computer that can run a basic communications and terminal emulator package + modem
Fees:	None – FREE
For more information:	*Dial-up*: 0203 362560/364436
Comments:	Yes, it's free but it is very busy.

ukmailNET/Infocom Interactive
(formerly DGI-Infocom)

Address:	The Davinson Group International, ukmail NETWORK, White Bridge House, Old Bath Road, CHARVIL, RG10 9QJ
POPs:	Reading
Proposed POPs (by end 1994):	
Basic services:	E-mail/USENET News from host machine (Infocom Interactive) or from your machine/site via UUCP (ukmailNET)
Additional services:	Freeware and shareware UUCP software for Acorn, Amiga, Atari, IBM, Mac and UNIX Systems. Teletext pages.
Help/Support:	1 month free, £25/year extra
Method of connection:	Dial-up
Form of service delivery:	Terminal access to BB service or via UUCP
Documentation:	
Minimum hardware/ software requirements:	Any computer that can run a basic communications and terminal emulator package or UUCP (+ modem)
Fees:	Single user (UUCP): Mail – £35/year, Mail/News – £50/year Multi-user (UUCP) Mail/News: From £290/year Infocom Interactive access: £45/year or £10 if combined with UUCP. (+VAT)
For more information:	_Tel_: 0734 344000 (Office hours)
	Tel: 0850 920041 (Out of office hours)
	Fax: 0734 320988
	E-mail: info.admin@gate.ukmail.net info@gate.ukmail.net (auto reply service, send empty note with 'Subject: ALL')
	Dial-up: 0734 340055 (login using 'fax' and follow instructions for Fax delivery of info.
Comments:	Currently a UUCP based service but may be soon moving to provide an interactive service based on SLIP/PPP.

WinNET Mail & News

Address:	PC User Group Ltd, PO Box 360, 84-88 Pinner Road, HARROW HA1 4LQ
POPs:	London
Proposed POPs (by end 1994):	
Basic services:	E-mail and USENET News on your machine
Additional services:	FTP, telnet and other Internet Services (Gopher, and so on) also available on host machine (at extra cost)
Help/Support:	
Method of connection:	Dial-up
Form of service delivery:	Proprietary communications package which automatically down/up-loads e-mail and USENET News + terminal access to other services
Documentation:	
Minimum hardware/ software requirements:	PC, Windows 3.1 and modem
Fees:	E-mail and USENET News Service – £3.25/hour connect (min £6.75 per month). FTP + telnet service – extra £7.25/month (no additional time charges on this part of the service) (+VAT)
For more information:	*Tel*: 081 863 1191 *Fax*: 081 863 6095 *E-mail*: request@win-uk.net *FTP*: ftp.ibmpcug.co.uk:/pub/doc/services *Dial-up*: 081 863 6646 (logon 'winnet' and follow instructions for a copy of the WINnet software)
Comments:	A service offered by the PC User Group. There is a separate entry for the PC User Group's Connect Service.

Part 3

Appendices

Some sorted lists

List 1 – Points of Presence by City

Below is a list of cities where access points or POPs can be found, with the name of the company or companies concerned.

Notes

* Where the POP is not in place at time of writing but is proposed by the end of 1994 the company name is in italics.

* Local rates may be available from nearby towns and cities, not just within the city listed, so it is worth checking for nearby cities in addition to the one you may be in.

* Some companies may charge extra for accessing a POP remote from their main site.

* Some UK companies offer X.25 access via GNS DialPlus. DialPlus has approximately 130 POPs all over the UK, thus giving local call access to most towns and cities. Telecom Eireann offers a similar service known as Eirpac. Access via these services may still be limited to 2400 baud in some areas in the UK but the majority of POPs will offer 9600 baud plus V.42bis or MNP5. DialPlus will also offer 14.4 Kbaud sometime in 1995.

Aberdeen
Edex, *ElectricMail, EUnet GB*

Belfast
ElectricMail, EUnet GB, Genesis Project

Birmingham
BBC Networking Club, CityScape, CompuServe, Demon, ElectricMail, EUnet GB, *Motiv Systems,* Pavilion, PIPEX, RedNet onLine

Bracknell
ElectricMail, EUnet GB

Bradford
Demon, Motiv Systems

Brighton
Pavilion

Bristol
BBC Networking Club, CityScape, CompuServe, *ElectricMail, EUnet GB*, Pavilion, PIPEX, RedNet onLine

Byfleet
Specialix International

Cambridge
BBC Networking Club, CityScape, *Demon*, ElectricMail, EUnet GB, *Motiv Systems*, PIPEX, RedNet onLine

Canterbury
ElectricMail, EUnet GB

Cork
Genesis Project, HEAnet, IEunet, INFONET, *Ireland On-Line*

Coventry
Spud's Xanadu

Dublin
Genesis Project, HEAnet, IEunet, Ireland On-Line

Edinburgh
BBC Networking Club, CityScape, CompuServe, Demon, ExNet, Motiv Systems, Pavilion, PIPEX, RedNet onLine

Galway
HEAnet, IEunet, Ireland On-Line

Glasgow
ElectricMail, EUnet GB

Grangemouth
ALMAC BBS

Hull
Demon, Motiv Systems

Kirklees
Kirklees Host

Leeds
Demon, Edex, Motiv Systems

Limerick
HEAnet, Ireland On-Line

London
BBC Networking Club, British Telecom, CityScape, CIX, CompuServe, Delphi, Demon, Direct Connection, Direct Line, Edex, ElectricMail, EUnet GB, ExNet, GreenNet, London Host, Motiv Systems, On-Line Entertainment, Pavilion, PC User Group, PIPEX, Poptel, RedNet onLine, Sprint International, WinNET

Manchester
BBC Networking Club, British Telecom, CityScape, CompuServe, Manchester Host, Pavilion, PIPEX, RedNet onLine

Portsmouth
Pavilion

Reading
CompuServe, Demon, Motiv Systems, ukmailNET

Shannon
IEunet

Swindon
ElectricMail, EUnet GB

Sheffield
Demon, Motiv Systems

Southampton
EUnet GB

Sunderland
Demon, Motiv Systems

Weybridge
Sound & Vision

Warrington
Demon, Motiv Systems

List 2 – Company by Form of Service Delivery

The following lists show which companies provide specific forms of service delivery (as defined in Chapter 3). Companies that provide or enable the use of off-line readers are also listed separately as some of the most recently launched services are based around this approach (WinNET and CityScape). Many companies have more than one form of service delivery available.

Notes

- As a full TCP/IP service over a leased line may be of particular interest to commercial readers, this is listed separately.

- Where a form of service delivery is not currently available from a company but that company has indicated it intends to provide it soon the company name is in italics.

Terminal access

ALMAC BBS, BBC Networking Club, CIX, CompuServe, Delphi, Demon, Direct Connection, Direct Line, Edex, ExNet, GreenNet, INFONET, Ireland On-Line, Kirklees Host, London Host, Manchester Host, Motiv Systems, On-Line Entertainment, PC User Group (CONNECT), Poptel, Sound & Vision, Spud's Xanadu, ukmailNET

Off-line reader(s) (Option)

ALMAC BBS, CIX, *Delphi,* Direct Line, INFONET, Ireland On-Line

Off-line reader (Part of service package)

CityScape, CompuServe, Sound & Vision, WinNET

UUCP batch delivery

Direct Connection, Edex, ElectricMail, EUnet GB, ExNet, IEunet, Ireland On-Line, PC User Group (CONNECT), Specialix International, ukmailNET

IP to your machine or LAN (Dial-up or on-demand links)

BBC Networking Club, British Telecom, Demon, Direct Connection, Edex, ElectricMail, EUnet GB, ExNet, Genesis Project, HEAnet, IEunet, Ireland On-Line, JANET (as a 'secondary' service), Motiv Systems, Pavilion, PIPEX, RedNet onLine

IP to your machine or LAN (leased line)

British Telecom, Demon, ElectricMail, EUnet GB, ExNet, Genesis Project, HEAnet, IEunet, Ireland On-Line, JANET, Motiv Systems, PIPEX Sprint International

---

 See also Appendix B – Possible major newcomers to Internet access. Here you will find contact details for the major telecommunications companies who may soon be offering Internet related services, aimed mainly at the corporate market.

List 3 – Company by Method of Connection

The following list shows which companies provide specific methods of connection. As explained in Chapter 3, the method of connection does not dictate the form of service delivery but some combinations are more sensible than others. For example, it might be possible to supply an IP service to a group of machines on a LAN via a high-speed modem and dial-up line but if this was in constant use it would probably be more realistic to go for a leased line plus router solution.

Notes

- If a company has indicated a connection method is possible but does not list it as part of standard service, the company name is in italics.

- JANET (the UK academic network) allows its members to link other organisations (assuming they pass the Acceptable Use Guidelines) via 'secondary' connections. In this case the method of connection will depend on the organisation providing the 'secondary' connection.

  Dial-up via modem
  ALMAC BBS, BBC Networking Club, CityScape, CIX, CompuServe, Delphi, Demon, Direct Connection, Direct Line, Edex, ElectricMail, EUnet GB, ExNet, Genesis Project, GreenNet, IEunet, INFONET, Ireland On-Line, Kirklees Host, London Host, Manchester Host, Motiv Systems, On-line Entertainment, Pavilion, PC User Group, PIPEX, Poptel, RedNet onLine, Sound & Vision, Specialix International, ukmailNET, WinNet

  Dial-up via modem to dedicated/reserved modem
  British Telecom, Demon, Genesis Project, Motiv Systems

  Dial-up via ISDN
  ALMAC BBS, British Telecom, CIX, ElectricMail, EUnet GB, *ExNet*, Genesis Project, PIPEX

  Direct or indirect (via dial-up service) X.25 access
  CIX, CompuServe, Delphi, Edex, ElectricMail, EUnet GB, GreenNet, Kirklees Host, London Host, Manchester Host, On-line Entertainment, PC User Group, PIPEX, Poptel

  Dial-up X.25 access (using the access provider's account)
  CompuServe, Edex, GreenNet, On-line Entertainment

  IP over X.25
  ElectricMail, EUnet GB

  Leased line
  British Telecom, Demon, ElectricMail, EUnet GB, ExNet, Genesis Project, HEAnet, IEunet, *Ireland On-Line*, JANET, Motiv Systems, PIPEX, Sprint International

  Leased line and router
  British Telecom, Demon, ElectricMail, EUnet GB, Genesis Project, IEunet, *Ireland On-Line*, JANET, Motiv Systems, PIPEX

  Internet/telnet
  CIX, On-Line Entertainment, PC User Group

B

Possible major newcomers to Internet access

The following major communications companies have either announced, or indicated in a more informal way, their intention (or probable intention) to offer Internet related or IP services in the UK or Republic of Ireland in the near future (late 1994, early 1995). In most cases, it has proved difficult to get detailed information as most of the companies concerned are in the pre-launch planning phase and do not want to give their competitors advance details of their new services. Most of these companies are mainly (or only) interested in corporate customers. If you are in this category they should be happy to talk to you even if they have not formally launched their new service.

AT&T – Business Communications

Contact: General Enquiries – Business Communications – Europe
Tel: 0527 514514
Status:
IP service already available in the US but no plans currently for IP in UK. This may change so check with General Enquiries.

British Telecom

Contact: Tony Sweet
Tel: 0442 237353
Fax: 0442 237082
Status:
Launch of IP and possibly other Internet services announced for November 1994.
Further details available in main Internet Access Providers listing – see British Telecom entry.

France Telecom

Contact: Les Shaw, Director of Sales and Marketing
Tel: 071 379 4747
Fax: 071 379 1404
Status:
France Telecom already provide Internet connectivity via Transpac (X.25) in France. At a recent conference (Internet World International '94, London, May 1994) Les Shaw outlined proposals for a Europe-wide service. No details of possible UK service or launch date yet.

Mercury Communications Ltd

Contact: Gary Muchmore (Mercury Messaging)
Tel: 081 914 6230
Fax: 081 014 6682
Status:
At the internal discussion stage. Launch date not known.

Sprint International

Contact: Robin Loxley
Tel: 0256 811181
Fax: 0256 811149
Status:
IP service (SprintLink) already available in the US. Worldwide Global SprintLink Service (GSS) being launched late 1994, including a node in London.
Further details available in main Internet Access Providers listing – see **Sprint International**.

Modems – what they are and what they do

Modem – another technical term

A modem is a device that connects your computer to the telephone network. It seems mandatory when mentioning modems in polite company to say that the name is short for 'modulator/demodulator', which is obvious to those who know and means nothing to those who do not! Quite simply a modem converts the signals put out by your computer (designed to be understood by another computer) into signals that can be carried by the telephone system, and it converts similar signals coming in over the telephone line back into signals your computer understands. A little thought indicates that modems must operate in pairs – one at each end of the call. This is both blindingly obvious and very important. The operative word is 'pairs'. All modems bought in the UK operate according to internationally agreed standards (the CCITT 'V' series) so they can all 'talk' to others that speak the same 'language' or standard. The standards define the ability to operate at different speeds (or offer different 'bandwidth' connections) and other important features (described below). The usual rule is that the faster a modem can operate the more it costs. High-speed modems can usually communicate with slow-speed modems *but only at the slower speed*. All modern modems have the ability to negotiate with each other as they connect, in order to to decide the fastest speed at which they can work together (this can be overridden but unless there is a good reason *and* you know what you are doing let them decide). In addition, most modern modems (other than the most basic) come with extra features like error correction and data compression. There are two series of standards for these operations, those from CCITT (parts of the 'V' series mentioned above), and those produced by the Microcom company (MNP series) which are so widespread (and available on other modems) that they have been adopted as industry (and in some cases international) standards. It is common for high-specification modems to have both series available. Finally, almost all modern modems are 'Hayes compatible' – this means they understand a simple control language invented by the Hayes modem company. All the currently available communications packages assume they are talking to a Hayes compatible modem.

In addition to coming in a range of speeds and with a range of features, modems also come in a range of physical forms. These include: the traditional flat box designed to fit under a telephone (these models are usually mains powered), a wide range of smaller boxes[1] (especially those designed to accompany portable computers – some powered from the computer and some battery powered), printed circuit boards (usually called 'internal modems' but sometimes confusingly called 'modem cards'[2]) for insertion in one of the standard PC's internal slots, and the new PCMCIA[3] modems which are the same size and shape as standard credit cards but vary in thickness[4]. These fit into the PCMCIA card-slots becoming available on the latest portables and some desktops.

As with modern computers, you don't need to understand all this to use a modem. When buying a modem the best rule of thumb is to buy the highest specification you can afford. It may be faster than your access provider currently supports but it will still work at the slower speed and when they upgrade you will be ready.

Notes

1   I've even seen one shaped like a circular bar of soap!

2   Originally this was OK as all add-in PC printed circuit boards are called 'cards', for example, serial card, video card – but now we have PCMCIA cards which complicates matters. The easiest way to distinguish between them is: a plug-in modem for a PC on a standard printed circuit board is a 'modem-card', and a modem on a PCMCIA card is a 'card-modem'.

3   Personal Computer Memory Card International Association.

4   Currently there are three 'types' of PCMCIA card: Type I is 3.3mm thick and is used mainly for memory applications, Type II is 5mm thick and is used for modem and LAN interface applications, and Type III is 10.5mm and is used for miniature hard disks. The slots are downwards compatible, that is, a Type II slot can take a Type I card, and so on.

X.400 service
providers in the UK

AT&T Mail

Address:

AT&T EasyLink Services	AT&T Easylink Services
Bastwick Street	4 Moons Park
London	Burnt Meadow Road
EC1V 3PH	Redditch WORCS B98 9PA
Tel: 071 251 1577	*Tel*: 0527 514514
Fax: 071 253 0416	*Fax*: 0527 498805

Comments:

Includes Fax and Telex gateways, plus interaction with their EDI network.

BT Mailbox Service

Address:

Not known
Tel: 0800 800916
E-mail: S=Sales-Enquiries O=BT Messaging Service A=BT C=GB
Comments:
Provides access to X.400 and Dialcom (Telecom Gold, and so on) services.
Includes fax and Telex links.

Mercury MultiMessage X.400

Address:

Mercury Communications Ltd
Mercury Messaging Sales
Profile West
950 Great West Road
Brentford
Middlesex TW8 9DS
Tel: 081 914 2500
Fax: 081 914 2366

Comments:

There is also a 'cc:Mail' gateway service known as *cc:Mail-2-MultiMessage*.

SprintMail 400

Address:

Sprint International
Intec 3, Wade Road
Basingstoke
Hampshire
RG24 0NE
Tel: 0256 811181
Fax: 0256 811149

Comments:

Fax, Telex, and file transfer services plus *PC SprintMail* direct from a PC.

E

APC networks

Association for Progressive Communications Networks[1]

Global computer communications for the environment, human rights, development & peace

What is APC?

The Association for Progressive Communications (APC) is a worldwide partnership of member networks dedicated to providing low-cost computer communications services for individuals and non-governmental organizations (NGOs) working for environmental sustainability, universal human rights and social and economic justice.

APC enhances the effectiveness of local, indigenous organizations by encouraging expertise in computer networking. Problems such as ozone depletion, violation of human rights, deforestation, hunger and oppression need international solutions. Regional, national and global solutions will only emerge from efficient communications.

All APC members are independent organizations that retain full control of their network. Member networks pay a percentage of their income to the APC Secretariat to diversify the growth of the Association.

APC was the primary provider of telecommunications for NGOs at the 1992 Earth Summit in Brazil, and in 1993 at the United Nations World Conference on Human Rights in Austria. Planning is under way to provide similar services in preparation for the 1994 UN International Conference on Environment and Development in Cairo, the 1995 UN World Summit on Social Development in Copenhagen and the 1995 Fourth World Conference on Women in Beijing.

Services provided on APC networks

Electronic mail
Electronic mail can be exchanged with MCI mail, Internet, HandsNet, FidoNet, Dialcom, Bitnet, UUCP, GeoNet, CompuServe and most other academic or commercial networks. You can send and receive telex, and send fax messages directly from the APC Networks.

Electronic conferences
Both public and private conferences are supported. The public conferences include events calenders, newsletters, legislative alerts, press releases, and so on. Private conferences can be used to facilitate internal group decision making, information exchange, and so on.

Databases
These include alternative publications, bibliographies, grant-making foundations, Third World Resources, Greenpeace Press Releases, Agenda 21, Shortwave Radio Transcripts, United Nations Information Service, and Pesticide Information Service.

List of APC Networks

ALTERNEX, IBASE, Rua Vicente de Souza 29

22251-070 Rio de Janeiro, Brazil
Fax: +55 (21) 286-0541
E-mail: suporte@ax.apc.org

CHASQUE, Casilla Correo 1539,

Montevideo 11000, Uruguay
Tel: +598 (2) 496-192
Fax: +598 (2) 419-222
E-mail: apoyo@chasque.apc.org

COMLINK e.v., Emil-Meyer-Str. 20,

D-30165 Hannover, Germany
Tel +49 (511) 350-1573
E-mail: support@oln.comlink.apc.org

ECUANEX, 12 de Octubre 622, Of. 504,

Casilla 17-12-566, Quito, Ecuador
Tel: +593 (2) 528-716
Fax: +593 (2) 505-073
E-mail: intercom@ecuanex.apc.org

GLASNET, ulitsa Sadovaya-Chernograizskaya,

dom 4, Komnata 16, Third Floor, 107078 Moscow, Russia
Tel: +7 (095) 207-0704
E-mail: support@glas.apc.org

GLUK — GlasNet-Ukraine, 14b Metrologicheskaya str.,

Kiev, 252143, Ukraine
Tel: +7 (044) 266 9481
Fax: +7 (044) 266 9475
Email: support@gluk.apc.org

GREENNET, 4th Floor, 393-395 City Road,

London EC1V WE , UK.
Tel: +44 (71) 608-3040
Fax: +44 (71) 253-0801
E-mail: support@gn.apc.org

ANTENNA, Box 1513,

NL-6501 Nijmegen, Netherlands
Tel: +31 (80) 235-372
Fax: +31 (80) 236-798
Email: support@antenna.nl

HISTRIA, Ziherlova 43 61

Ljubljana, Slovenija
Tel: +38 (61) 211-553
Fax: +38 (61) 152-107

Email: support@histria.apc.org

IGC - EcoNet/PeaceNet/Conflict/LaborNet
18 De Boom Street, San Francisco, CA 94107, USA
Tel: +1 (415) 442-0220
Fax: +1 (415) 546-1794
E-mail: support@igc.apc.org

LANETA, Tlalpan 1025, col. portales,

Mexico, df., Mexico
Tel: +52 (5) 277-4791
Fax: +52 (5) 277-4791
Email: soporte@laneta.apc.org

NICARAO, CRIES, Apartado 3516, Iglesia Carmen,

1 cuadra al lago, Managua, Nicaragua
Tel: +505 (2) 621 312
Fax: +505 (2) 621 244
E-mail: ayuda@nicarao.apc.org

NORDNET, Huvudskaersvaegen 13, nb,

S-12154 Johanneshov, Sweden
Tel: +46 (8) 6000-331
Fax: +46 (8) 6000-443
E-mail: support@nn.apc.org

PEGASUS, PO Box 284,

Broadway 4006, Queensland, Australia
Tel: +61 (7) 257-1111
Fax: +61 (7) 257-1087
E-mail: support@peg.apc.org

SANGONET, 13th floor Longsbank Building,

187 Bree Street, Johannesberg 2000, South Africa
Tel: +27 (11) 838-6944
Fax: +27 (11) 838-6310
E-mail: support@wn.apc.org

WAMANI, CCI, Talachuano 325-3F,

1013 Buenos Aires, Argentina
Tel: +54 (1) 35 6842
E-mail: apoyo@wamani.apc.org

WEB, NirvCentre, 401 Richmond Street West,

Suite 104, Toronto, Ontario M5V 3A8, Canada
Tel: +1 (416) 596-0212
Fax: +1 (416) 596-1374
E-mail: support@web.apc.org

APC Secretariat Offices

APC International Secretariat,

Rua Vincente de Souza, 29,
22251-070 Rio de Janeiro, BRASIL
Tel: +55(21)286-4467
Fax: +55(21)286-0541
Email: apcadmin@apc.org

APC North American Regional Office,

18 De Boom Street,
San Francisco, CA 94107, USA
Tel: +1(415)442-0220
Fax: +1(415)546-1794
Email: apcadmin@apc.org

[1] This Appendix is a shortened and edited version of the _APC On-line Brochure_ (93/94) written by Edie Farwell and provided to the author by Viv Kendon of GreenNet.

Extended glossary

The Internet is a jargon-ridden place. The following is a selection of terms or phrases you will often find in articles or books about the Internet. Unless there is a special reason, the names of access-providing companies are not listed here.

In some cases a single word, phrase or even an acronym (for example *netiquette, backbone network* or *FAQ*) may hold a whole new set of ideas. For this reason, some of the explanations are far more than a simple definition.

 Italicised terms in the explanations are themselves defined either elsewhere in the Glossary or in the main text.

Acceptable Use Guidelines
The rules that govern the use of many academic networks, for example *NSFNET* in the US and *JANET* in the UK. They usually impose limitations on who may connect to them and what kind of things they may be used for.

Acceptable Use Policy
See Acceptable Use Guidelines.

Alias
Many *mailer* packages have a facility that allows you to associate a real name (or some easily remembered substitute) with that person's e-mail address. For example, if you have a colleague called Fred Jones with an address something like: XX236P1D@ somewhere_with_a_long_and_difficult_name.com' you can associate this address with the name 'fredj'; when you want to mail him you just need to put 'fredj' in the 'To:' field of the note and the mailer will replace it with the full address just before mailing. With some mailers the alias can also point to a list of addresses rather than a single address.

Anonymous FTP
A technique for making a vast amount of information and software available for public access. *See* Chapter 2 for details.

ARPANET (or ARPAnet)
Advanced Research Projects Agency Network. The predecessor of the Internet. The ARPANET was the original test-bed for the techniques underpining the Internet

Article
A posting in a *USENET Newsgroup.*

Archie
A filename locating service associated with *anonymous FTP.*

ASCII
American Standard Code for Information Interchange.

Backbone network
A network that links other (usually geographically smaller) networks. For example, *NSFNET* is the backbone network for the academic networks in the US.

Bandwidth
The data carrying capacity of a communications link. *See* also *Baud, bits per second* and Chapter 3.

BB
See *Bulletin Board.*

Baud
A measure of communications capacity or *bandwidth*. One *Baud* means one signalling element per second. In a signalling system with only two possible states (on/off), *Baud* and *bit per second* have the same meaning. See also Chapter 3, under Methods of connection.

Bit(s) per second
In a communications system, the standard measure of *bandwidth* or data transmission speed (not to be confused with propagation speed – *see end-to-end performance*). Usually used in multiples of 1000 (kilobit/second) or 1 million (megabit/second). In many cases, '/second' is left off, that is, it is common for the speed (or *bandwidth*) of a LAN to be expressed as '10 megabits' (or whatever) with no mention of the time period, because this is assumed to be one second by *default*. See also Chapter 3, under *Methods of connection*.

BITNET (Because It's Time Network)
A worldwide computer network based on NJE protocols (originally for IBM mainframes) supporting messaging, e-mail and file transfer (not Internet FTP). Strictly speaking, only the part that is in the USA and Mexico should be called BITNET as the worldwide network includes other national and international networks including EARN (European Academic and Research Network). However, there is no global name for these interconnected NJE based networks so the whole group is called by the name of its largest and oldest member – BITNET. See also *EARN*.

Bounce or bounced
The return of e-mail that could not be delivered. Usually accompanied by an error message and/or diagnostic information. Sometimes used to mean the sending of mail via a third site known to have extensive address tables for another network (see also *Mail Gateway*).

bps
See *bit(s) per second*.

BUBL (Bulletin Board for Libraries)
A Gopher-based information service of interest to the library and information community (and others) running on the UKOLN machine at the University of Bath. It was originally a 'traditional' *Bulletin Board* service, hence the title. BUBL is maintained mainly by voluntary effort (drawn from all over the UK via the network) and coordinated by Dennis Nicholson from the University of Strathclyde.

Bulletin Board
A common area on a computer where users can read or write generally visible messages. By having areas made accessible only by password, or other form of checking system, 'private' BBs can be set up for group discussions. Based on the idea of a paper-based public notice board

where messages can be 'pinned-up' for others to see and reply to. See also Chapter 2.

CIX (Commercial Internet eXchange)
A cooperative grouping of the major commercial IP network providers in the US and the rest of the world who have agreed to interwork their networks.

CIX (Compulink Information eXchange)
A company offering Internet access in the UK. See the Internet Access Providers list for details.

Client
See *Client/Server*

Client/Server
Instead of having one program on one machine working to provide a service like *telnet* or *FTP*, which requires an interaction between two machines, there are usually two complementary programs (one on each machine) that cooperate closely to achieve the desired result. The usual rule is that the program started by the user on the local machine which requests a service from a distant machine is called the *client*, and the program that provides the service is called the *server*. Again, because the system is designed to accommodate any two machines the two programs need to communicate in a standard (or common) form known as a *protocol*. See also Chapter 1 and Figure 1.2.

Compress
Text files often have a great deal of 'wasted' space in them; single characters, or patterns of characters, are repeated. These repetitions can be replaced by codes that represent the repeated patterns but take up less space. Using this approach (and other techniques) it is possible to compress a text file by up to half. This means it takes up less disk space and may be quicker to transfer. There are programs to do this on most computers: compress on UNIX, PKZIP on PCs, StuffIt on Macs, and so on. Obviously there are also 'uncompress' programs to reverse this. Each program is different, so a compressed file needs to be uncompressed using its exact opposite. For example, PKUNZIP cannot uncompress a file compressed with StuffIt.

CCITT (Comité Consultatif International Télégraphique et Téléphonique)
An international body (part of the ITU – International Telecommunications Union) that sets standards for telecommunications services that cross national boundaries. The origin of many standards in the world of telecommunications, including the 'X' series concerned with packet switching services and the 'V' series concerned with modem operation.

CSNET
Computer Science Network (academic network that once paralleled ARPANET).

Cyberspace
the universe of networked computers and all their programs and services. The term was made popular by William Gibson in his book *Neuromancer*.

Dedicated line
In this book, I have used *dedicated line* to mean a dedicated number (for the subscriber's use only) into the access provider's machine. This gives the same level of availability and access as a leased line (unless one of the exchanges between you and the access provider is full – rare in most of the UK), but at less cost if you only need intermittent access. I have used *dedicated line* in this way because I noticed that companies who offer this service tend to call it a dedicated or reserved line. However, I am aware that others use *dedicated line* and *leased line* to mean the same thing. See also *leased line*.

Default
If something is usually the same in all, or in the great majority of cases, it is often not mentioned as it is assumed by the writer or speaker to be understood by the reader or listener. See *bit(s) per second* for an example of this.

Default value
An extension of the *default* principle. If the value of a command parameter is usually the same, it is often arranged by the system designers or software writers that this value is used if you do not specify otherwise. See *port* or *port number* for an example of this.

Dial-up
This is access via a modem and telephone line or via ISDN. It differs from *dedicated line* access in that you cannot be sure that there will be a spare modem or ISDN connection at the service provider's end when you need it. *Dial-up* is by far the most common method of initial connection for individuals and small organisations connecting to Internet services.

Dial-in IP
This is where you run IP direct from your machine and link to your access provider's machine via a dial-up service, usually using a modem over a telephone line. See Full interactive (TCP/IP) access in Chapter 3.

Download

The act of transferring data or text from a distant computer to your local one. The opposite of *upload*. The transfer may be by *FTP* or *screen logging* or some other transfer technique.

E-Journal (Electronic Journal)

A range of electronic information provision services set up to mimic paper-based journals or magazines. There are various modes of delivery: an 'issue' may be distributed by *mailing list* and delivered as a single note via *e-mail*; the complete text of an 'issue' may be made available on an *FTP archive* while a 'contents page' is distributed by *e-mail* and *mailing list*; the individual articles that make up an 'issue' may be made available separately on the *FTP archive* and the contents distributed by *e-mail,* and so on. The permutations are large. In addition, the latest e-journals are making use of the *Web* to make graphics and other features available to those with an appropriate *Web browser.* Most *e-journals* are free but some require a subscription to be paid for access.

Electronic Journal

See *E-journal.*

E-mail (electronic mail)

Either (verb) to transfer a text message in a standard format file between users on the same or remote machines, or (noun) the file that is transferred. See Chapter 2.

EARN (European Academic and Research Network)

An NJE (IBM protocol) based network which has over 600 sites in 40 countries and stretches from Northern Europe to the Middle East and North Africa. See also *BITNET.*

EFF (Electronic Frontier Foundation)

A foundation formed to consider the possible social and legal impact of the Internet on society.

E-mail conference

See *Mailing list* and Chapter 2.

Elm

A full-screen based UNIX mail program. A great improvement over the standard UNIX mail program.

Emoticon

See *Smiley.*

End-to-end performance

When considering network performance, in addition to the capacity of the link, you may need to consider the actual time it takes for a signal to travel from you to a distant service and back again. This is especially true if you have an application waiting at your end which cannot proceed until it gets that response. The delay or turn-around time is usually expressed in milliseconds (ms). A service contract for supplying a connection to a commercial organisation will usually contain a clause specifying the average delay to be expected.

This is obviously partly related to _bandwidth_ and directly related to how fast the signal travels (the propagation speed). As an analogy think of a river: _bandwidth_ is the volume of water that flows past a given point during a given time period, while the propagation speed is the speed at which the water flows.

FAQ (Frequently Asked Questions)

The same basic questions tend to be asked over and over again by new users of e-mail lists or USENET News Newsgroups. After a time someone will collect these questions and answers (sometimes summarised) together and post them regularly in the appropriate lists or News-groups. If you are intending to ask a question of a list or Newsgroup it is considered good _netiquette_ to check for any FAQ file first and if there is one, check it out before asking.

FidoNet

A worldwide network based on dial-up PCs and other micro-computers running the FidoNet Bulletin Board package. FidoNet reaches parts of the world other networks fail to reach (to coin a phrase). For example, there are FidoNet networks in Africa and parts of the Far East where the Internet is still only a rumour. FidoNet currently has over 20,000 nodes (each with many users) and is growing at around 30% a year. It is possible to e-mail FidoNet users from the Internet. E-mail _Hostmaster@f1.n1.z31.fidonet.org_ for further information.

Flame

Because of the informal tone of e-mail and/or the Newsgroups, the ease of quick response, and the fact that the other person is not physically present, arguments can sometimes break out during discussions (especially over sensitive topics like religion or politics). These strongly worded notes are called flames. Sometimes these exchanges can grow and overflow into other related lists and Newsgroups and become flame-wars. Best avoided unless you enjoy arguments, know how to create a _kill-file_ and have an asbestos-covered ego. If you are tempted to flame – stop, go outside and kick something inanimate (but not breakable), count up to one hundred – then reply if you must. See also _netiquette_.

Follow-up

A reply to a _USENET News_ article or _posting_.

Freeware
Software made available by the author at no cost. Not quite the same as Public Domain Software (PDS) as the author still retains copyright. See also *Shareware*.

Frequently Asked Questions
See *FAQ*.

Front-end
Either the user interface part of an application program, or a separate program, that sits between the user and an unfriendly operating system or application program. The front-end may provide a menu to the user and pass appropriate commands to the application when a menu item is selected.

FTP (File Transfer Protocol)
See Chapter 2.

FTP archive
See Chapter 2.

FYI (For Your Information)
The initials FYI are often used to precede a reply to a question or a comment on an ongoing discussion where the sender does not expect a reply.

Gateway
A computer that sits on or between two networks that use different underlying protocols and enables communication between them. Sometimes dedicated to one specific service, for example, a *mail gateway*.

Gopher (or Internet Gopher)
A *client/server* based network navigation tool providing a menu-like interface. See Chapter 2 for more details.

Gopherspace
All the *Gopher* servers on the network.

ISDN (Integrated Services Digital Network)
A digital network that supports both voice and data. An ISDN connection replaces the existing analogue telephone connection – instead you have an ISDN box to which you can connect true ISDN equipment (including telephones, high-speed faxes and ISDN cards for PCs) and in addition, if the connecting adaptor allows for it, analogue equipment (ordinary telephones, and so on). The basic ISDN service (ISDN 2) provides two 64 Kbit/sec data channels and a 16 Kbit/sec control channel. Commercial organisations can opt for ISDN 30 which provides thirty 64 Kbit/sec channels. See also Chapter 3.

Header
When used in the context of IP packets, this means the bits encoding address information that precede the actual data. When used in the context of e-mail, it means the first part of the document that precedes the actual message text. It consists of the sender's and recipient's names and addresses, the subject, the date, and other routine information. See also the section on e-mail in Chapter 2.

Host
A computer on the Internet that you dial-in to for Internet services if you have *terminal* access (or a *log-in account*). If you have permanent IP connectivity your machine is a host, or if you have *dial-in IP* your machine is a host for the time it is connected.

HTML (Hyper Text Markup Language)
The scripting language used to mark up *Web* documents. Enables the links followed by *Web browsers*.

Hypermedia
A combination of *hypertext* and *multimedia*.

Hypertext
The underlying idea behind the Web. Hypertext documents contain embedded pointers to other text that adds to, or expands on, the text currently being read. This text may be part of the current document or another document. See also the introduction to the section on WWW in Chapter 2.

Internet Gopher
See *Gopher.*

IP connection
This means the IP data packets are sent from, and delivered to, your machine. See Full interactive (TCP/IP) access, Chapter 3.

IRC (Internet Radio Chat)
See Chapter 5, under Additional services.

JANET (Joint Academic Network)
The network that links all UK academic and government research laboratory sites in the UK. Strictly speaking, JANET is a multi-protocol network (including X.25 and IP services) and only the JIPS (JANET IP Service) is part of the Internet. JANET has over one million users spread over 200+ sites and includes over 100,000 individually addressable machines.

JIPS
See *JANET.*

Kermit

This is both a general-purpose file transfer protocol and a terminal emulation package that implements the protocol.

Kill-file

A list of senders' names or subjects of e-mail notes (or Newsgroup articles) that you *do not* want to see. Any incoming mail or UNSENT News article that matches the criteria in the applicable kill-file is thrown away before you see it. Be careful in defining the criteria, if you are too general you will lose items you may have wanted. Useful if you get involved in *flaming* or otherwise are being sent mail you do not want to be bothered with. This can also happen if you are on a mailing list that in general interests you, but a topic is being discussed that bores you stiff. The exact way in which you set up a kill-file will depend on the mail package or *News-reader* that you use – check the documentation.

LAN (Local Area Network)

A communications network confined to a single building or site.

Leased line

In the 'olden days' this might have been a genuine dedicated telephone wire connection from your premises to those of the access provider. Today, with most long-distance communications going over optical fibre or microwave links, it is really a guaranteed connection with a given bandwidth from you to the access provider (although the final connection at each end will probably be a real wire link to the local exchange). See also *dedicated line.*

LISTSERV

This is an automated mailing list manager package designed originally for large IBM machines linked via *BITNET.* There are now thousands of mailing lists on every imaginable topic maintained by LISTSERVs around the world. A LISTSERV may maintain tens or hundreds of lists on one machine. Although it is a large package with many features, its main job is to allow remote users to join or leave *mailing lists* automatically by sending commands or instructions as text in an e-mail note – hence you don't need to be on *BITNET* to use a LISTSERV mailing list. In addition, it can respond to requests for help and other information, maintain archives of correspondence on the lists it manages and offer full-text searching of these archives. It has to be confessed that the techniques you have to learn to search a LISTSERV -maintained archive are far from friendly but they do work.

Note: A common problem for new users of automated mailing lists is the confusion between the list-maintaining program and the lists maintained. To send something to the list you use an address of the form *listname@host* (where host may be a BITNET address or an Internet address), whereas to send a command to the LISTSERV program you use an address of the form LISTSERV@host. So to join a list called TOLKIEN (which is a discussion list concerned with the works of J R R Tolkien) maintained by a LISTSERV at the site JHUVM on BITNET you would send an e-mail note to LISTSERV@JUHVM.bitnet containing the following line:

SUBSCRIBE TOLKIEN your_firstname your_lastname.

To send a note to be distributed to the subscribers to TOLKIEN you would send it to **TOLKIEN@JHUVM.bitnet.**

Log-in account
See *Terminal access* in Chapter 3.

Lurker
Someone who often reads contributions in a *USENET Newsgroup* but does not contribute. Despite the rather derogatory name this is perfectly acceptable behaviour (lurkers probably outnumber contributors to most Newsgroups by 10 or 20 to 1). If you don't have something useful to say, don't say it – it saves network resources!

Mail gateway
A machine that sits on two networks with different e-mail protocols (that is, with different forms of addressing and/or internal e-mail file structure). It translates addresses and mail files as they pass through, enabling transparent e-mail exchange.

Mailer
The program you interact with when sending or receiving e-mail. See also *Elm* and *Pine*.

Mailing list
This is a list of e-mail addresses which you can mail to as if it were a single address. The list may be private or public. If it is public, it may also be called a discussion list or an e-mail conference or a mail-reflector. The most common use of such a service is for the discussion of a topic of mutual interest to those on the list. Each note sent to the list is an 'open letter' to all those on the list and so are all the replies. The early lists were maintained by a human operator who added or removed addresses on request. This was superseded for most lists by programs (for example *LISTSERV*) that accept commands via incoming e-mail notes and add or delete addresses from the lists they maintain. See *LISTSERV* for an overview of the operation of a popular mailing list maintaining program. See also Chapter 2.

MAN (Metropolitan Area Network)
A communications network confined to a town, city or large area of a city.

MNP (Microcom Network Protocol)
A series of standards for error correction (MNP1-4) and data compression (MNP5), originally from the Microcom modem company but now adopted as industry and international standards. Most modern modems incorporate both the *CCITT* V.42 (error correction) and V.42bis (data compression) standards and the MNP standards.

MODEM
A piece of hardware that enables a computer or other digitally based equipment to communicate across telephone lines which were designed for analogue communications. See Appendix C.

Mosaic
A family of *Web browser* clients for PCs, Macs and other machines, written and distributed by the NCSA (National Center for Supercomputing Applications).

MIME (Multipurpose Internet Mail Extensions)
A standard that allows e-mail users to include non-text files (programs in binary form, image files, and so on.) in an e-mail note. Only works automatically if the mail handling programs at both ends of the transfer understand it. Mainly uses existing techniques for transferring non-text in e-mail form (for example, *uuencode/uudecode*) but automates the coding and decoding process. This means that if you do not have a MIME compliant mailer you may still be able to decode the resultant file embedded in the incoming mail if you know what you are doing and you have the appropriate tools (software).

Multimedia
Documents that contain a range of data. For example, some *Web* servers have information services that contain a combination of text, graphics, moving images and sound. Some CD-ROM based encyclopedias (for example) are multimedia.

Netiquette
A compound of the words 'net' and 'etiquette'. A set of tacitly agreed rules concerning the way you should behave when using the network. Like etiquette in society, these are not laws but generally understood guidelines. Not everyone will agree with the following and others would add many more, but these are my suggestions:
• Just because your e-mail system allows you to send notes of any length, do not embed the text of your latest book on some topic in a note sent in response to a general enquiry. The recipient may not be able to receive such a large note, may not have your privileges

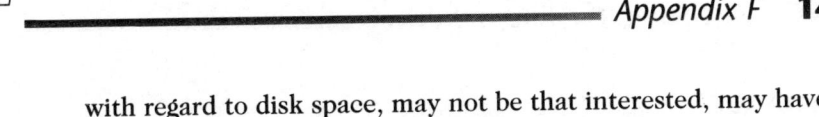

with regard to disk space, may not be that interested, may have to pay for incoming e-mail, may have to make an expensive long distance call just to access their mail, and so on.

- Think before you reply to something that annoys you (*see Flames*). Would your boss, neighbour, spouse, mother, friends, and so on, be happy with your response – will you be happy with it next week, month, year? Remember, even if you delete it from your filestore, it will still be out there on tens, hundreds or thousands of machines (especially on USENET).

- When retrieving large files, think about your impact on the network. Even if you really want it, is this the best time to retrieve it – could it wait until the network is less busy? Remember the net is worldwide, so if you want a large file from the US, retrieve it when the US is asleep (when it is morning in the UK).

- Respect the privacy of others. If someone sends you an interesting (or annoying) note privately, do not reveal its contents or distribute it to a mailing list without asking if it's OK with the sender first.

- Remember the net is worldwide and there are many using it with different cultural backgrounds from your own. Don't assume they will understand your ironic or sarcastic comments. Be especially careful with humour when there is a need for a shared history or cultural background. Take into account others' ethical or religious views – even if you do not agree with them.

- Finally, there are many net users for whom English is a second, third or fourth language – be tolerant of grammar and spelling. How well do you speak Swahili?

Newsgroup

See *USENET Newsgroup*.

News-reader

This is a package that enables you to read *USENET News* articles. It may run on the access provider's machine if you have *terminal access,* or on your machine if you have *full I/P* access or *batch access*. See Chapter 2 for more information on USENET and Chapter 3 for forms of access.

NREN (National Research and Education Network)

A major initiative in the US to provide network access to all those in education, including schools and colleges of all sizes. Currently at the research and implementation stage. Seen by some as the precursor to the fabled 'Information Superhighway'.

NSFNET

National Science Foundation Network – the backbone network for academic networks in the US.

Packet switching

The underlying communications technique of the Internet. See Chapter 1.

Paste or pasting

This refers to the ability of many communications and terminal emulation programs to paste or write information to the screen (and the distant computer) from a file – as if it were being typed at the keyboard. This can be very useful if you use a host based log-in account for e-mail. By preparing the text of an e-mail note beforehand, logging in and opening the distant e-mail editor to receive text, and then pasting the ready-prepared text into the body of the note, you can save a lot of on-line time (and probably use a much more friendly text editor as well!).

Pine

A UNIX mail program similar to *Elm* but with a menu driven *front-end* and its own editor.

Port or Port Number

In hardware terms 'port' refers to one of the sockets on the back or side of your machine. When used as part of the address of a service on a remote machine it refers to a particular application running on that machine. For example, Gopher servers are usually on Port 70, so you do not need to specify the port number when connecting to a standard Gopher server (this is known as a *default value*). Where there is more than one Gopher server on a machine, they are distinguished by having different port numbers. For example, we have two generally accessible Gopher servers on the UKOLN machine at Bath. To distinguish between them, the second one has ' Port 7070' after the address.

Post

Putting an article into a *USENET Newsgroup*.

Protocol

A set of rules concerning how two computers 'talk' to each other. This may be at a very low level like the Internet Protocol (IP) which underpins all communication across the Internet, or it may be a set of rules that define the messages sent between application-level programs like the *Client/Server* pairs that provide such services as Gopher.

PPP (Point-to-Point Protocol)

Provides IP over a serial line connection (as provided by a modem and telephone line). Similar to SLIP but written later and considered technically superior by many users.

Relevance feedback

A feature of *WAIS*. See Chapter 2, under WAIS.

Router

Hardware that connects two networks with similar protocols but that are physically different. For example, a *LAN* and the Internet.

Screen logging

This refers to the ability of many communications and terminal emulation programs to copy text from the screen as it comes in from a distant computer to a file on your machine. This can be a quick and easy way to download small to medium text files or copies of your e-mail for later reading. Especially useful if you have a fast modem with error correction and data compression.

Server

See *Client/Server*.

Signature file

A file containing information like the sender's name and address and e-mail address that is automatically added to the bottom of all outgoing e-mail. It is considered bad *netiquette* to have a very large signature-file, especially if you usually send very short notes.

Shareware

Software made freely available for evaluation. If you intend to continue to use the software you should pay a fee to the author who will usually then provide the latest version and some support. See also *Freeware*.

Shell

In UNIX a program that sits between you and the operating system, interpreting your commands and shielding you from the rather user-hostile interface of naked UNIX. In some cases *terminal access* or *log-in account* services are called *shell accounts*.

SLIP (Serial Line Internet Protocol)

A protocol for providing IP over a dial-up telephone link or similar communications medium. See also *PPP.*

Smiley (or Smileys)

Because it is difficult in a short e-mail note to convey emotions that would qualify what is being said, small 'pictures' formed mainly from punctuation marks have been evolved to add this extra dimension.

The simplest is the *smiley* to indicate humour. Here are three examples:

:-) Indicates humour, or "I didn't mean exactly what I just wrote – take it as ironic" (or similar).

:-("It makes me sad", or "I'm sad".

;-) A winky or "nudge, nudge – know what I mean?" sort of comment.

These are just a few examples – the possibilities are endless. Popular articles, academic papers and even a small book (*Smileys* by D W Sanderson, O'Reilly Associates, ISBN 1-56592-041-4) have been written on the rise of the smiley (or to use its techie name – *emoticon*). This symbolic representation of emotions may be, dare I say it, a sign of the times ;-) .

telnet

The Internet protocol for remote access or remote login. See Chapter 2.

Terminal access

See Chapter 3.

tn3270

A telnet-like protocol used to access large IBM machines (over the Internet) which expect connections from IBM3270 full-screen terminals rather than *VT100* terminals.

Upload

The act of transferring data or text from your local computer to a distant one. The opposite of *download*. The transfer may be by FTP, *Kermit,* or some other form of standard file transfer technique, or by *pasting*.

USENET

User Network. Supports USENET News – *see* Chapter 2.

USENET News

A distributed form of *Bulletin Board*. See Chapter 2.

USENET Newsgroup

That part of USENET News dedicated to the discussion of a particular subject, similar to a *mailing-list* or *e-mail* conference.

UUCP (Unix to Unix Copy Program)

A basic utility program in UNIX that enables file transfer between UNIX based computers.

UUCPnet

A collection of UNIX computers all over the world that exchange e-mail and other files using the *UUCP* file transfer protocol. UUCPnet pre-dates the Internet, but it now uses the Internet where available to provide the underlying connectivity between distant machines. Like the Internet, there is no single network; many loosely interconnected networks use UUCP internally and externally as their communication standard. The main international e-mail routing centre is at UUNET in the US. For more information e-mail *info@uunet.uu.net.*

uudecode
See *uuencode.*

uuencode
A UNIX program to convert binary files to ASCII files. Why? – because with text-only e-mail systems (still the majority) you cannot transfer a binary image or program file (or a word-processor file with embedded format characters) in the body of a note, as some of the bit patterns could be interpreted as control codes by the transfer programs. To get around this, uuencode converts binary files to ASCII character (that is, text only) files which can be transferred in the body of an e-mail note. Uuencoded files are approximately 30–40% larger than the binary files they represent. To convert the files back you use the *uudecode* program. If you receive a note where the body part starts:

begin <some-number> <some-text>

followed on the next line by a solid block of nonsense ASCII characters it has probably been uuencoded.

Veronica–
A keyword search service associated with *Gopher.* See Chapter 2 under *Gopher.*

V series
These are the *CCITT* series of recommendations for 'data transmission over telephone lines', that is, they are the international standards for this area. They are usually seen used in conjunction with modems to define their operating speeds and other features. The most common ones seen today are:

V.22bis	= 2400 bps
V.32	= 9600 bps
V.32bis	= 14400 bps
V.42	= Error correction standard
V.42bis	= Data compression standard

VT100
Originally the product name of one of Digital Equipment Corporation's terminal families. An early 'full-screen' terminal where the application

driving it could write to or read from selected areas of the whole screen, rather than just the last line as was the case with the original scrolling screen VDUs. Now it is the most emulated terminal standard and is used as the 'lowest common denominator' for access to the great majority of on-line services.

W3

See *World Wide Web* – Chapter 2.

WAIS (Wide Area Information Server)

A *client/server* based network navigation tool for full-text sources. See under *WAIS* in Chapter 2 for more details.

Web

See *World Wide Web* – Chapter 2.

Web browser

In order to interact with a *Web* server you need a *Web client*, which are also known as *Web browsers*. The better known packages are Mosaic and Cello for both PCs (with Windows) and Macs, and Lynx on UNIX for those with VT100 (or similar) access. See also Chapter 2.

World Wide Web

See Chapter 2.

WWW

See *World Wide Web* – Chapter 2.

X.25

Strictly speaking, this is the *CCITT* standard defining the interface between DTE (Data Terminal Equipment) and DCE (Data Circuit-termination Equipment) operating in packet switched mode. It is now used as 'shorthand' to mean packet switched data networks that operate according to the CCITT standards.

G

Suggested further reading

The following is a list of recent books in this area, with comments. In addition, there is one book concerned with modems and their operation which may be of use to those new to dial-up services.

The books are categorised according to whether they are *user guides, specialist, reference, or technical*. In some cases this is a difficult decision and the choice made may appear arbitrary. However, there is a sort of logic at work here. User guides are for those who have some knowledge of computers, may have access or are seriously considering access and want to know more about the Internet. Specialist books are aimed mainly at those who have a connection, or have decided they want a connection, and are focused on some specific topic, such as, e-mail, getting connected, a specific type of user, and so on. Reference books are as their name suggests – books not to be read from cover to cover but to be referred to as needed. The only technical book listed is the one concerned with modems and their use.

Thereafter the books are listed alphabetically by title.

The *Rating* after the *Comments* attempts to rank the value of the book within its category. *Useful* means it is well worth a look, while *above average* and *excellent* mean exactly what they say.

 None of these books are aimed at communications engineers, or similar specialists – they are all aimed at general readers with some computer knowledge.

User guides

How to use the Internet
Mark Butler
145 pages.
Ziff-Davis Press 1994
ISBN 1-56276-222-2
£16.49
Comments:
This is a large format 'picture' book on the Internet and related topics, for example, the basics of UNIX and computing in general. The focus is on terminal access to a log-in account. The colour graphics are strong and well drawn. In many places the old adage about a 'picture is worth a thousand words' works very well and some ideas that are difficult to express in words are clear as a picture. However, there are times when the exuberance of the artist definitely gets in the way of the message.

There is a US bias which is particularly obvious in the discussion of e-mail. There are many pictures of US street mailboxes and those strange tin boxes on sticks which always look to me like aircraft hangars for pigeons but sit at the end of the drives to US homes for the delivery of paper mail.

Each topic is given a two-page spread and the text paragraphs are numbered clockwise across both pages (an approach I found confusing). Despite my reservations, if you find that pictures and diagrams make technical ideas easier to absorb this might be just the Internet book for you – certainly worth a look.
Rating: Useful

The Internet Guide for New Users
Daniel Dern
570 pages
McGraw-Hill 1993
ISBN 0-07-016511-4
£22.95
Comments:
This is a big book in many ways; it weighs over two pounds, is nearly one and a half inches thick and covers a very wide range of topics – encyclopedic is the word that comes to mind! All the usual topics are covered, from ARPANET to ZIP compression, via e-mail, FTP, USENET and the rest. For example, over 40 pages are used to describe the basics of UNIX – useful if you have, or intend to have, a log-in account on a UNIX system or your local machine is UNIX based. The style is friendly, if sometimes verbose, and analogies are often used to good effect when explaining the technology. There are many references by name to real people on the net, which adds to the feeling of friendliness (and gives an insight to the network 'community'). The Contents 'page' is eleven pages long! However, even with this almost page-by-page description (and a quite comprehensive Index), it can be difficult to find some things due to the amount of information contained in this book.
Rating: Above average

Internet Starter Kit for Windows
Adam Engst, Corwin Low and Michael Simon
608 pages
Hayden Books 1994
1-56830-094-8
£27.98
(Note: There is also a Mac version of this book – *Internet Starter Kit for the Macintosh* ISBN 1-56830-064-6)
Comments:
This is the biggest so far – nearly three pounds in weight and nearly two inches thick! However, looks can be deceiving – part of the thickness is the disk inside the back cover (of which more later). Also, nearly 250 pages are devoted to listings of Internet resources and a further 23 pages to an elaborate Glossary. The space taken by these two sections plus some appendices leaves less than half the book to be a conventional (but very technically oriented) introduction to the Internet.
However, this is more than a book – it is a book with real software. Included on the disk is a cut-down copy of Chameleon from NetManage. This contains WinSock and TCP/IP software to work with a SLIP connection. Also included is Telnet, TN3270, FTP, Ping and

Mail. In addition, there is a copy of the Eudora e-mail package from QUALCOMM, WinVN (an NNTP News reader) and WinSock Gopher. In order to use all this you need a SLIP account and, were you to be in the US, a special deal has been set up for you to access such a service from Northwest Nexus for a two-week free trial. Unfortunately, for you to use this from the UK would involve you in rather expensive trans-Atlantic calls! Still – if you are serious about connecting and you want SLIP based access to a UK host, you have all the basic tools here for a very realistic price (and you can upgrade at a later date to the full Chameleon package if you want). The book is written for the computer buff who is happy to tweak various bits of software to get them to work together. If you are happy to experiment and have some idea what you are doing, this combination of book and software (plus a UK SLIP service provider) could be just what you are looking for. It really is a self-contained kit.

Rating: Above average

The Complete Idiot's Guide to the Internet
Peter Kent
386 pages
Alpha Books 1994
ISBN 1-56761-414-0
£18.50

Comments:

Aimed at the total beginner, the style is light and even flippant in places with a sprinkling of cartoons to lighten the message even more. Each Chapter is short and focused. The book follows the classic training style – beginning each Chapter with a list of the topics to be covered, and ending it with a summary. The small boxes for tips and techno-speak are similar in style to the *Internet for Dummies* book (*see* below), but this is more lightweight. There is also a form of glossary ('Speak like a geek' archive) which is more than a simple list of the terms in the boxes spread throughout the book. In addition to the text in the book, there is a disk inside the back cover containing self-extracting archives which expand to text files with lists of mailing lists, USENET Newsgroups, and so on. All of this is available on the Internet but for those who do not yet have access, it could provide an insight into some of the resources on the network. The blurb on the front cover talks of an 'e-mail utility' but the only operational program is a UUENCODE / DECODE program which could be used to include a binary file in a basic e-mail note. Strangely, although UUENCODE /DECODE is discussed in Chapter 12, that information is not included in the Index. This may indicate a problem with the construction of the Index.

Rating: Useful

The Internet Navigator
Paul Gilster
470 pages
Wiley 1993
ISBN 0-471-59782-1
£21.95

Comments:
A very well-written and thought-out introduction to the Internet aimed specifically at the dial-up user. The author starts by describing the sorts of things that can be done with access to the Internet – before he goes on to explain the technical ideas. This is a good approach – you are given a reason to plough through what follows even if you are not a techie. The style is somewhere between that of Daniel Dern and the SRI team (below), friendly but crisp. Occasionally he goes into too much detail on specific programs (like the UNIX Mail program) that you may never use. The coverage of USENET and Gopher is particularly good but the coverage of the Web is disappointing (probably due to when the book was written – the Web has grown enormously in the past year).
Rating: Above average

STOP PRESS:
There is now a 2nd edition of this book.

The Internet for Dummies
John Levine & Carol Baroudi
355 pages
IDG Books 1993
ISBN 1-56884-024-1
£17.99
Comments:
One of the range of *for Dummies* books. Similar to the *Complete Idiot's Guide to the Internet* reviewed above but slightly less chatty in style and with more solid technical information in the separate text boxes. Few cartoons (unless you count the small icons used to indicate different types of information) and a smaller typeface mean there is more information in this book than in the *Complete Idiot's Guide...*, despite the smaller number of pages. With six parts broken into twenty-nine chapters, each chapter is even more bite-sized than in the *Complete Idiot's Guide*. This book and the *Complete Idiot's Guide* are obviously in direct competition. In terms of solid information this one has the edge – but the *Complete Idiot's Guide* has the disk....
Rating: Above average

Internet: Getting Started
April Marine, et al
359 pages
PTR Prentice Hall (SRI Internet Information Series) 1993
ISBN 0-13-327933-2
£22.50

Comments:

A general introduction to the Internet. Written by a small team of four, the style is almost exactly opposite to Daniel Dern (*see* above), being dry and antiseptic. This can be useful – especially when you just want the facts about a topic. It includes an international list of access providers with contact information, plus many useful lists and addresses of Internet related organisations. Similar in some ways to the book you are reading, but with an international and US bias.

Rating: Useful

The Whole Internet – User's Guide & Catalog (2nd ed.)
Ed Krol
543 pages
O'Reilly & Associates 1994
ISBN 1-56592-063-5
£18.50

Comments:

This is the second edition of what many believe to be the best user's guide to the Internet. It is not just the coverage that makes it so good (Daniel Dern probably covers more, and Paul Gilster at least as much) – it is the style of writing and clarity of thought. Rarely are technical books a pleasure to read, but this one is. All the basic stuff is here: the history of the network, e-mail, FTP, News, Gopher, WAIS, WWW, and so on, plus a short resource catalog (his spelling). The new edition is over 40% larger than the previous one but the price is the same. Although all the sections are a little larger, the main areas of growth are the sections on the Web (over twice as long), the list of resources in the Catalog (nearly twice as long) and the appendix on getting connected (again over twice as long). The amount of thought that has gone into this book is indicated by the fact that there are two indexes, one for the first half of the book (the Guide) and another for the second half (the Catalog). Highly recommended.

Rating: Excellent

Specialist

Connecting to the Internet
Susan Estrada
170 pages
O'Reilly & Associates 1993
ISBN 1-56592-061-9
£11.95
Comments:
In many ways similar to this book (but with a US bias). Aimed slightly
more at the commercial/professional user. More attention is paid to
such things as network performance (speed, reliability, security, and
so on). Worth consulting just for the check-lists she provides to help
choose an access provider.
Rating: Above average

!%@:: A Directory of Electronic Mail Addressing and Networks
Donnalyn Frey & Rick Adams
443 pages
O'Reilly & Associates 1993 (3rd Ed)
ISBN 1-56592-031-7
£20.50
Comments:
The bulk of the book consists of a list of the main networks worldwide,
each with a one-or two-page description. Excellent but not recommended for the beginner (except Chapter 1 which explains E-mail to as
great a level of detail as you are ever likely to need). This book has
been available since 1989 and is in its third edition – which, in this
rapidly changing area, says it all. Bedtime reading for Internet gurus!
Rating: Excellent

Internet Primer for Information Professionals
Elizabeth Lane and Graig Summerhill
181 pages
Meckler 1993
ISBN 0-88736-831-X
£21.00
Comments:
Similar in style to the *Internet: Getting started* from SRI but with a
leaning towards the library and information professional. Although it
is an introduction to the Internet in general, it occasionally delves
quite deeply into the underlying technology. This could be of value to
some Technical Services librarians. Although it has a strong US bias,

Chapter Six (Policy Issues) will be of particular interest to librarians and information specialists. The three-page bibliography at the end clearly shows the authors' library and information science background.
Rating: Useful

Reference

Internet World's on Internet 94
Tony Abbott (ed.)
(with an Introduction by Daniel Dern)
453 pages
Mecklermedia 1994
ISBN 0-88736-929-4
ISSN 1066-9973
£24.00
Comments:
Returning to the theme of judging books by their size and weight, this one is the biggest in this list. It's three pounds in weight and the size of a telephone directory – which is not surprising as it is a *directory*. It is a directory or guide to what is on the Internet – E-Journals, Newsletters, E-mail (discussion) lists, USENET Newsgroups, and so on. This is a brave attempt to summarise just what is available. Brave because there is so much, and brave because it changes day by day (probably minute by minute). Despite the dynamism of the Internet's resources, a printed version is still of value (though it will date quickly) and this is the best I've seen. Many of the introductory books listed above have lists of resources (amounting to half the book in the case of the *Internet Starter Kit*) but none are as comprehensive and well indexed as this. Used in conjunction with one of the beginners' guides, this will give any reader a real feel for what is out there. As the title includes a date and an ISSN, I expect annual updates.
Rating: Above average

Technical

Modem and Communications Guidebook
  Sue Schofield
  345 pages
  Future Publishing Ltd 1993
  ISBN 1-85870-000-0
  £19.95

Comments:
  Although this is not directly concerned with connecting to the Internet (though there are a few pages on CompuServe, CIX and the Internet), most people making their first connection to the Internet will do so via a modem or ISDN (covered in Chapter 11). For some, this may be their first encounter with the strange world of telecommunications and they may feel threatened by such terms as *9600 baud, V.32bis,* and so on. This book should help. It contains far more information on modems and communications than is likely to be needed by the average single user, but is probably essential for someone considering the options for a small company or similar organisation. For the lone user its presence on the shelf could be very useful if things don't work as expected.

Rating: Above average

Index

Notes

Notes

Notes

Notes

Notes

Notes

Notes

Books from

International Thomson Publishing

On The Internet

PIECING TOGETHER MOSAIC
Navigating the Internet and the World Wide Web
Steve Bowbrick, 3W Magazine

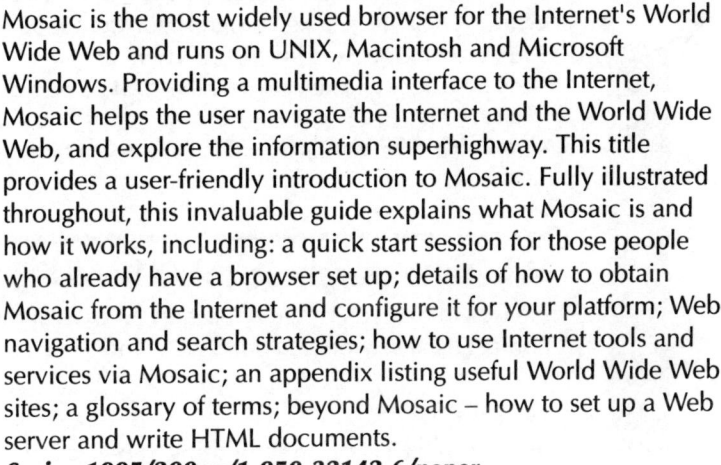

Mosaic is the most widely used browser for the Internet's World
Wide Web and runs on UNIX, Macintosh and Microsoft
Windows. Providing a multimedia interface to the Internet,
Mosaic helps the user navigate the Internet and the World Wide
Web, and explore the information superhighway. This title
provides a user-friendly introduction to Mosaic. Fully illustrated
throughout, this invaluable guide explains what Mosaic is and
how it works, including: a quick start session for those people
who already have a browser set up; details of how to obtain
Mosaic from the Internet and configure it for your platform; Web
navigation and search strategies; how to use Internet tools and
services via Mosaic; an appendix listing useful World Wide Web
sites; a glossary of terms; beyond Mosaic – how to set up a Web
server and write HTML documents.
Spring 1995/300pp/1-850-32142-6/paper

SPINNING THE WEB
How to Provide Information on the Internet
Andrew Ford

An indispensable guide for all those who provide or intend to
provide information on the World Wide Web, or want to make
the most of their existing services, this book for the first time
draws together all of the most up to date information and details
of contemporary resources into one essential volume. Providing
exclusive coverage of Web features, the book includes an
overview of Web facilities, how to create hypertext documents,
security issues, how to set up a server and the selection and
evaluation of software. A variety of examples from current Web
sources are included.
December 1994/250pp/1-850-32141-8/paper

On CompuServe

COMPUSERVE FOR EUROPE
Roelf Sluman

CompuServe, the world's largest personal on-line service, allows access to a world of information and services – plus a gateway to the Internet, the information super highway. News, financial reports, hobbies, travel, entertainment, interest groups, forums and electronic mail are just a few of the range of services available on-line via CompuServe. Written with the European user in mind, this is the ideal guide to this on-line service. Whether an existing member or a first-time user, it provides help and advice in a readable, accessible way. It also provides a WinCIM disk free, a key program for CompuServe access – plus $15* credit for new and existing users.
*CompuServe is an international service and is priced in \$US. Billing is in local currency at the prevailing rate.
December 1994/448pp/1-850-32121-3/paper

Where to purchase these books?
Please contact your local bookshop, in case of difficulties, contact us at one of the addresses below -

ORDERS
International Thomson Publishing Services Ltd
Cheriton House, North Way, Andover, Hants SP10 5BE, UK
Telephone: 0264 332424/Giro Account No: 2096919/
Fax: 0264 364418

SALES AND MARKETING ENQUIRIES

International Thomson Publishing
Berkshire House, 168/173 High Holborn, London WCIV 7AA,UK
Tel: 071-497 1422 Fax: 071-497 1426
e–mail: Info@ITPUK.CO.UK

MAILING LIST
To receive further information on our Networks books, please send the following information to the London address –
Full name and address (including Postcode)
Telephone, Fax Numbers and e-mail address

INTERNET

Books from O Reilly & Associates

The Whole Internet User's Guide & Catalog
By Ed Krol
2nd Edition, April 1994
574pages, ISBN 1- 56592-063-5

The best book about the Internet just got better! This is the second edition of our comprehensive – and bestselling – introduction to the Internet, the international network that includes virtually every major computer site in the world. In addition to email, file transfer, remote login, and network news, this book pays special attention to some new tools for helping you find information. Useful to beginners and veterans alike, this book will help you explore what's possible on the Net. Also includes a pull-out quick-reference card.

"An ongoing classic."
– *Rochester Business Journal*

"The book against which all subsequent Internet guides are measured, Krol's work has emerged as an indispensable reference to beginners and seasoned travelers alike as they venture out on the data highway."
— *Microtimes*

"*The Whole Internet User's Guide Catalog* will probably become the Internet user's bible because it provides comprehensive, easy instructions for those who want to get the most from this valuable electronic tool."
— David J. Buerger, Editor, *Communications Week*

"Krol's work is comprehensive and lucid, an overview which presents network basics in clear and understandable language. I consider it essential."
— Paul Gilster, *Triad Business News*

!%@:: A Directory of Electronic Mail Addressing & Networks
By Donnalyn Frey & Rick Adams
4th Edition June 1994, 662 pages. ISBN 1-56592-046-5

This is the only up-to-date directory that charts the networks that make up the Internet, provides contact names and addresses, and describes the services each network provides. It includes all of the major Internet-based networks, as well as various commercial networks such as CompuServe, Delphi, and America Online that are "gatewayed" to the Internet for transfer of electronic mail and other services. If you are someone who wants to connect to the Internet, or someone who already is connected but wants concise, up-to-date information on many of the world's networks, check out this book.

"The book remains the bible of electronic messaging today. One could easily borrow the American Express slogan with the quip 'don't do messaging without it.' The book introduces you to electronic mail in all its many forms and flavors, tells you about the networks throughout the world... with an up-to-date summary of information on each, plus handy references such as all the world's subdomains. The husband-wife team authors are among the most knowledgeable people in the Internet world. This is one of those publications for which you just enter a lifetime subscription." – Book Review, *ISOC News*

The Mosaic Handbooks

Mosaic is an important application that is becoming instrumental in the growth of the Internet. These books, one for Microsoft Windows, one for the X Window System, and one for the Macintosh, introduce you to Mosaic and its use in navigating and finding information on the World Wide Web. They show you how to use Mosaic to replace some of the traditional Internet functions like FTP, Gopher, Archie, Veronica, and WAIS. For more advanced users, the books describe how to add external viewers to Mosaic (allowing it to display many additional file types) and how to customize the Mosaic interface, such as screen elements, colors, and fonts. The Microsoft and Macintosh versions come with a copy of Mosaic on a floppy disk; the X Window version comes with a CD-ROM.

The Mosaic Handbook for Microsoft Windows
By Dale Dougherty and Richard Koman
Ist Edition October 1994, 234 pages. ISBN 1-56592-094-5 (Floppy disk included)

The Mosaic Handbook for the X Window System
By Dale Dougherty, Richard Koman and Paula Ferguson
Ist Edition, October 1994, 220 pages, ISBN 1-56592-095-3 (CD-ROM included)

The Mosaic Handbook for the Macintosh
By Dale Dougherty & Richard Koman
Ist Edition October 1994 , 220 pages, ISBN 1-56592-096-1 (Floppy disk included)

Connecting to the Internet

By Susan Estrada
1st Edition, August 1993
188 pages, ISBN 1-56592-061-9

This book provides practical advice on how to get an Internet connection. It describes how to assess your needs to determine the kind of Internet service that is best for you and how to find a local access provider and evaluate the services they offer.

Knowing how to purchase the right kind of Internet access can help you save money and avoid a lot of frustration. This book is the fastest way for you to learn how to get on the Internet. Then you can begin exploring one of the world's most valuable resources.

"A much needed 'how to do it' for anyone interested in getting Internet connectivity and using it as part of their organization or enterprise. The sections are simple and straightforward. If you want to know how to connect your organization, get this book."
– Book Review, *ISOC News*

Learning the UNIX Operating System

By Grace Todino, John Strang & Jerry Peek
3rd Edition, August 1993
108 pages, ISBN I-56592-060-0

If you are new to UNIX, this concise introduction will tell you just what you need to get started and no more. Why wade through a 600-page book when you can begin working productively in a matter of minutes? It's an ideal primer for Mac and PC users of the Internet who need to know a little bit about UNIX on the systems they visit. This book is the most effective introduction to UNIX in print. The third edition has been updated and expanded to provide increased coverage of window systems and networking. It's a handy book for someone just starting with UNIX, as well as someone who encounters a UNIX system as a "visitor" via remote login over the Internet.

If you have someone on your site who has never worked on a UNIX system and who needs a quick how-to, Nutshell has the right booklet. *Learning the UNIX Operating System* can get a newcomer rolling in a single session. It covers logging in and out; files and directories; mail; pipes; filters; background-ing; and a large number of other topics. It's clear, cheap, and can render a newcomer productive in a few hours." – *;login*

Smileys

By David W Sanderson
1st Edition March 1993, 93 pages, ISBN 1-56592-041-4

"For a quick grin at an odd moment, this is a nice pocket book to carry around :-) If you keep this book near your terminal, you could express many heretofore hidden feelings your email ;-) Then again, such things may be frowned upon at your company :-(No matter, this is a fun book to have around."
– Gregory M. Amov, *News & Review*

TCP/IP Network Administration

By Craig Hunt
1st Edition August 1992, 502 pages ISBN 0-937175-82-X

TCP/IP Network Administration is a complete guide to setting up and running a TCP/IP network for administrators of networks of systems or lone home systems that access the Internet. It starts with the fundamentals: what the protocols do and how they work, how to request a network address and a name (the forms needed are included in an appendix), and how to set up your network. Beyond basic setup, the book discusses how to configure important network applications, including sendmail, the r* commands, and some simple setups for NIS and NFS. There are also chapters on troubleshooting and security. In addition, this book covers several important packages that are available from the Net (such as *gated*). Covers BSD and System V TCP/IP implementations.

Managing Internet Information Services

By Cricket Liu, Jerry Peek, Russ Jones, Bryan Buus & Adrian Nye
1st Edition Winter 1994/95 (est), 400 pages, ISBN 1-56592-062-7

This comprehensive guide describes how to set up information services to make them available over the Internet. Providing complete coverage of all popular services, it discusses why a company would want to offer Internet services and how to select which ones to provide. Most of the book describes how to set up email services and FTP, Gopher, and World Wide Web servers.

"*Managing Internet Information Services* has long been needed in the Internet community, as well as in many organizations with IP-based networks. Although many on the Internet are quite savvy when it comes to administering these types of tools, MIIS will allow a much larger community to join in and perhaps provide more diverse information. This book will be a welcome addition to my Internet shelf."
– Robert H'obbes' Zakon, MITRE Corporation

sendmail

By Bryan Costales, with Eric Allman & Neil Rickert
1st Edition November 1993, 830 pages, ISBN 1-56592-056-2

Although sendmail is used on almost every UNIX system, it's one of the last great uncharted territories – and most difficult utilities to learn – in UNIX system administration. This book provides a complete sendmail tutorial, plus extensive reference material. It covers the BSD, UIUC IDA, and VR versions of sendmail.

"The program and its rule description file, sendmail.cf, have long been regarded as the pit of coals that separated the mild Unix system administrators from the real fire walkers. Now, sendmail syntax, testing, hidden rules, and other mysteries are revealed. Costales, Allman, and Rickert are the indisputable authorities to do the text."
– Ben Smith, *Byte*

DNS and BIND

By Cricket Liu & Paul Albitz
lst Edition October 1992, 418 pages ISBN 1-56592-010-4

DNS and BIND contains all you need to know about the Internet's Domain Name System (DNS) and the Berkeley Internet Name Domain (BIND), its UNIX implementation. The Domain Name System is the Internet's "phone book"; it's a database that tracks important information (in particular, names and addresses) for every computer on the Internet. If you're a system administrator, this book will show you how to set up and maintain the DNS software on your network.

"At 380 pages it blows away easily any vendor supplied information, and because it has an extensive troubleshooting section (using nslookup) it should never be far from your desk – especially when things on your network start to go awry :-)"
– Ian Hoyle, BHP Research, Melbourne Laboratories

MH & xmh: E-mail for Users & Programmers

By Jerry Peek
2nd Edition Septetnber 1992, 728 pages, ISBN 1-56592-027-9

Customizing your email environment can save time and make communicating more enjoyable. *MH & xmh: E-Mail for Users & Programmers* explains how to use, customize, and program with the MH electronic mail commands available on virtually any UNIX system. The handbook also covers *xmh,* an X Window System client that runs MH programs. The second edition added a chapter on mhook, sections explaining under-appreciated small commands and features, and more examples showing how to use MH to handle common situations.

Practical UNIX Security

By Simson Garfinkel & Gene Spafford
1st Edition June 1991, 512 pages, ISBN 0-937175-72-2

Practical UNIX Security tells system administrators how to make their UNIX system either - System V or BSD - as secure as it possibly can be without going to trusted system technology. The book describes UNIX concepts and how they enforce security, tells how to defend against and handle security breaches, and explains network security (including UUCP, NFS, Kerberos, and firewall machines) in detail. If you are a UNIX system administrator or user who deals with security, you need this book.

"Timely, accurate, written by recognized experts... covers every imaginable topic relating to Unix security. An excellent book and I recommend it as a valuable addition to any system administrator's or computer site manager's collection."
– Jon Wright, *Informatics (Australia)*

Where to purchase these books?

Please contact your local bookshop, in case of difficulties contact us at one of the addresses below -

ORDERS
International Thomson Publishing Services Ltd
Cheriton House, North Way, Andover, Hants SP10 5BE, UK
Telephone: 01264 332424/Giro Account No: 2096919/
Fax: 01264 364418
Email: UK orders - ITPUK@ITPS.CO.UK
        Outside UK orders - ITPINT@ITPS.CO.UK

SALES AND MARKETING ENQUIRIES
International Thomson Publishing
Berkshire House, 168/173 High Holborn, London WCIV 7AA, UK
Tel: 0171-497 1422 Fax: 0171-497 1426
e-mail: Info@ITPUK.CO.UK

MAILING LIST
To receive further information on our Networks books, please send the following information to the London address -
Full name and address (including Postcode)
Telephone, Fax Numbers and e-mail address